Kleinian Groups which Are Limits of Geometrically Finite Groups

MEMOIRS
of the
American Mathematical Society

Number 834

Kleinian Groups which Are Limits of Geometrically Finite Groups

Ken'ichi Ohshika

September 2005 • Volume 177 • Number 834 (second of 4 numbers) • ISSN 0065-9266

American Mathematical Society
Providence, Rhode Island

2000 *Mathematics Subject Classification.*
Primary 57M50, 30F40.

Library of Congress Cataloging-in-Publication Data

Ohshika, Ken'ichi, 1961–
Kleinian groups which are limits of geometrically finite groups / Kenichi Ohshika.
p. cm. — (Memoirs of the American Mathematical Society, ISSN 0065-9266 ; no. 834)
"Volume 177, number 834 (second of 4 numbers)."
Includes bibliographical references and index.
ISBN 0-8218-3772-9 (alk. paper)
1. Kleinian groups. 2. Low-dimensional topology. 3. Geometry, Hyperbolic. I. Title. II. Series.

QA3.A57 no. 834
[QA331]
510 s—dc22
[514′.22] 2005048026

Memoirs of the American Mathematical Society

This journal is devoted entirely to research in pure and applied mathematics.

Subscription information. The 2005 subscription begins with volume 173 and consists of six mailings, each containing one or more numbers. Subscription prices for 2005 are $606 list, $485 institutional member. A late charge of 10% of the subscription price will be imposed on orders received from nonmembers after January 1 of the subscription year. Subscribers outside the United States and India must pay a postage surcharge of $31; subscribers in India must pay a postage surcharge of $43. Expedited delivery to destinations in North America $35; elsewhere $130. Each number may be ordered separately; *please specify number* when ordering an individual number. For prices and titles of recently released numbers, see the New Publications sections of the *Notices of the American Mathematical Society.*

Back number information. For back issues see the *AMS Catalog of Publications.*

Subscriptions and orders should be addressed to the American Mathematical Society, P. O. Box 845904, Boston, MA 02284-5904, USA. *All orders must be accompanied by payment.* Other correspondence should be addressed to 201 Charles Street, Providence, RI 02904-2294, USA.

Memoirs of the American Mathematical Society is published bimonthly (each volume consisting usually of more than one number) by the American Mathematical Society at 201 Charles Street, Providence, RI 02904-2294, USA. Periodicals postage paid at Providence, RI. Postmaster: Send address changes to Memoirs, American Mathematical Society, 201 Charles Street, Providence, RI 02904-2294, USA.

This publication is indexed in *Science Citation Index®*, *SciSearch®*, *Research Alert®*, *CompuMath Citation Index®*, *Current Contents®/Physical, Chemical & Earth Sciences.*
Printed in the United States of America.

♾ The paper used in this book is acid-free and falls within the guidelines
established to ensure permanence and durability.
Visit the AMS home page at http://www.ams.org/

10 9 8 7 6 5 4 3 2 1 10 09 08 07 06 05

Contents

Abstract

Ahlfors conjectured in 1964 that the limit set of every finitely generated Kleinian group either has Lebesgue measure 0 or is the entire S^2. We prove that this conjecture is true for purely loxodromic Kleinian groups which are algebraic limits of geometrically finite groups. What we directly prove is that if a purely loxodromic Kleinian group Γ is an algebraic limit of geometrically finite groups and the limit set Λ_Γ is not the entire S^2_∞, then Γ is topologically (and geometrically) tame, that is, there is a compact 3-manifold whose interior is homeomorphic to $\mathbf{H}^3/\Gamma$. The proof uses techniques of hyperbolic geometry considerably and is based on works of Maskit, Thurston, Bonahon, Otal, and Canary.

AMS Subject Classfication: 57M50, 30F40

Keywords and phrases: Kleinian group, Ahlfors' conjecture, Marden's conjecture, topological tameness

Introduction

In 1964, Ahlfors raised a fundamental conjecture on finitely generated Kleinian groups as follows, which has not been solved yet.

CONJECTURE 1 (Ahlfors [**1**]). Let Γ be a finitely generated Kleinian group. Then its limit set in the Riemann sphere S^2 either has Lebesgue measure 0 or is equal to the entire sphere.

Although several partial solutions have been given to this Ahlfors' conjecture, among which is work of Ahlfors himself proving that the conjecture is true for the geometrically finite Kleinian groups, the conjecture itself is still open. From topological point of view, Ahlfors' conjecture is closely related to structures of ends of hyperbolic 3-manifolds, as was revealed by Thurston's pioneering work in his lecture notes [**45**]. Thurston introduced there the concept of geometrically tame Kleinian groups, which involves both topological and geometric structures of neighbourhoods of ends. According to Thurston, a Kleinian group Γ is said to be geometrically tame if each end of the non-cuspidal part of the hyperbolic 3-manifold $\mathbf{H}^3/\Gamma$ is either geometrically finite or simply degenerate in the following sense. An end is called geometrically finite when it has a neighbourhood intersecting no closed geodesics; and is called simply degenerate when it has a neighbourhood homeomorphic to the product of a surface and $\mathbf{R}$, and there exists a sequence of pleated surfaces tending to the end, on which there are no compressing curves whose lengths go to 0. This latter definition will be explained more precisely later.

Bonahon proved, in his celebrated paper [**5**], that indeed only the topological condition is necessary for hyperbolic 3-manifolds to be geometrically tame: that is, if the non-cuspidal part of $\mathbf{H}^3/\Gamma$ has a relative compact core each of whose frontier component is incompressible, then Γ is geometrically tame, in particular, that Ahlfors' conjecture is true for such Kleinian groups. This means that purely topological condition is sufficient to ensure a Kleinian group to be geometrically tame. It should be noted that in the case when a Kleinian group has no parabolic elements, this condition is equivalent to one that the group admits no non-trivial free product decomposition.

On the other hand, Canary proved that even for a hyperbolic 3-manifold M whose relative compact core is boundary-reducible, if each end of M has

Received by the editor February 3, 2000

a neighbourhood homeomorphic to the product of a surface and an open interval, then M is geometrically tame; in particular that Ahlfors' conjecture is true for such a manifold.

Thus Ahlfors' conjecture was proved to be derived from the following conjecture due to Marden:

CONJECTURE 2 (Marden [**25**]). Let Γ be a finitely generated torsion-free Kleinian group. Then, $\mathbf{H}^3/\Gamma$ is almost compact, i.e., there exists a compact 3-manifold whose interior is homeomorphic to $\mathbf{H}^3/\Gamma$.

This conjecture was motivated by one of the main theorems of [**25**], where he proved that this is the case for geometrically finite Kleinian groups.

Since geometrically finite Kleinian groups are far well-understood than geometrically infinite ones, and also all the geometrically infinite groups constructed thus far are algebraic limits of quasi-conformal deformations of geometrically finite groups, it is natural to consider the conjectures first for such limit groups. Indeed, Thurston conjectured in [**46**] the following, which is a generalization of Bers' conjecture on b-groups appeared in [**3**]:

CONJECTURE 3 (Thurston). Every finitely generated Kleinian group is an algebraic limit of quasi-conformal deformations of a geometrically finite group.

The goal of this paper is to prove that for non-free Kleinian groups without parabolic elements, with non-empty domain of discontinuity, which are algebraic limits of quasi-conformal deformations of geometrically finite Kleinian groups, Marden's conjecture is true. This will be proved by showing in such a situation the algebraic convergence of quasi-conformal deformations is also geometric. This result implies our main theorem stating that Ahlfors' conjecture is true for the Kleinian groups without parabolic elements which are algebraic limits of quasi-conformal deformations of geometrically finite Kleinian groups (Theorem 2.1), since Ahlfors' conjecture is known to be true for purely loxodromic free Kleinian groups by Maskit [**26**], and if the domain of discontinuity is empty the conjecture trivially holds. Our main theorem is regarded as reducing Ahlfors' conjecture to Thurston's conjecture above for Kleinian groups without parabolic elements.

The proof of the main theorem roughly consists of four steps. In the first step (chap.4), we shall consider a generalization of Thurston's argument which was used to prove that Ahlfors' conjecture is true for geometrically tame Kleinian groups in the original sense. We shall deal with hyperbolic 3-manifolds without cusps that contain compact cores homeomorphic to compression bodies. Such a hyperbolic 3-manifold has only one end that faces a compressible boundary component S of a compact core. Based upon an idea of Otal [**38**], we shall define geometric tameness for such an end in terms of simple closed curves on the boundary component S, which can be homotoped to closed geodesics going to the end, using a subspace called the Masur domain of the measured lamination space of S. We shall prove that if a hyperbolic 3-manifold has a compact core that is a compression body, and the end facing the compressible boundary component of a compact core has a property which is a generalization of geometric tameness, then Ahlfors' conjecture is true for the corresponding Kleinian group. In the meantime, we shall also show that such a hyperbolic 3-manifold is almost compact.

In the second step (chap. 6), we shall consider a sequence of quasi-conformally conjugate geometrically finite Kleinian groups $\{\Gamma_i\}$ converging strongly, viz., both algebraically and geometrically, to a geometrically infinite Kleinian group Γ' without parabolic elements such that $\mathbf{H}^3/\Gamma'$ has a compact core homeomorphic to a compression body. We shall prove that such a limit group satisfies the condition given in the first step, chap. 4, which is a generalization of geometric tameness.

In the third step (chap. 7), we shall consider a sequence of quasi-conformal deformations $\{\Gamma_i\}$ of a geometrically finite Kleinian group without parabolic elements converging algebraically to a Kleinian group Γ' without parabolic elements such that $\mathbf{H}^3/\Gamma_i$ has a compact core homeomorphic to a compression body. We shall prove that then either the limit set of Γ' is the entire sphere, or the convergence is strong and $\mathbf{H}^3/\Gamma'$ has a compact core homeomorphic to a compression body. This means that except for the case when the limit set of Γ' is the entire sphere, we can apply the result in the second step, chap. 6, hence also that in the first step, chap. 4, for such a limit.

In the final step (chap. 8), we shall consider quasi-conformal deformations $\{G_i\}$ of a geometrically finite Kleinian group without parabolic elements in general. We shall suppose that $\{G_i\}$ converges algebraically to a Kleinian group G without parabolic elements. Let Γ_i be a subgroup of G_i corresponding to the fundamental groups of a component of characteristic compression body of a compact core of $\mathbf{H}^3/G_i$. We shall show that the algebraic limit $\Gamma \subset G$ of $\{\Gamma_i\}$ also corresponds to the fundamental group of a component of a characteristic compression body of a compact core of $\mathbf{H}^3/G$.

Combining these four steps, the proof of our main theorem will be completed.

Canary and Minsky ([**11**]) have generalized our main theorem to one for purely loxodromic algebraic limits of topologically tame Kleinian groups based on a generalization of our argument in chap. 6. They have also proved independently a theorem similar to Theorem 6.1 in the present paper, which asserts that every purely loxodromic strong limit of topologically tame Kleinian groups is topologically tame.

Besides the conjecture with which we deal in this paper, Ahlfors also conjectured that if a finitely generated Kleinian group G has empty domain of discontinuity, then G acts on S^2_∞ ergodically. In relation to this type of Ahlfors' conjecture, we can ask if in the situation of our theorem, i.e., when G is purely loxodromic algebraic limit of quasi-conformal deformations of a geometrically finite group, the conjecture is true. It is proved by Canary [**7**] that this conjecture also follows from Marden's conjecture. Thus we see that the only missing part to prove the conjecture for limit groups as in our main theorem is the strong convergence in the case when G has empty region of discontinuity. As our argument in §7 essentially depends on the assumption that the group has non-empty region of discontinuity, it seems that we need a really new technique to fill this part.

Fairly large part of this work is done during the author's stay at Institut des Hautes Etudes Scientifiques. The author would like to express his gratitude to IHES for its hospitality and financial support. He would also like to thank Professor Bernard Maskit, who listened to my idea patiently and gave me valuable comments at the early stage of this work. The author is indebted to the referee for his/her valuable comments, which, the author hopes, has made the paper more readable.

CHAPTER 1

Preliminaries

In this section, we shall define notions and terms which will be used in this paper. We shall not go into details but cite references in which the reader can find further details.

1.A. Generalities

A Kleinian group is a discrete subgroup of the Lie group $PSL_2\mathbf{C}$. A Kleinian group with torsion has a torsion-free subgroup of finite index by Selberg's lemma [**41**]. Since the limit set, (which will be defined below,) of a finite index subgroup of a Kleinian group is the same as that of the original Kleinian group, we have only to deal with torsion-free Kleinian groups to consider Ahlfors' conjecture. Hence in this paper, we always assume that Kleinian groups are torsion free. Also several of results which we shall cite in this paper hold only for Kleinian groups without torsion.

The non-trivial elements of $PSL_2\mathbf{C}$ are classified into three types: the *loxodromic* elements which are conjugate (modulo $\pm I$) to matrices of the form $\begin{bmatrix} \lambda & 0 \\ 0 & \lambda^{-1} \end{bmatrix}$ with $|\lambda| > 1$; the *elliptic* elements which are conjugate to matrices of the form $\begin{bmatrix} \omega & 0 \\ 0 & \omega^{-1} \end{bmatrix}$ with $|\omega| = 1$; and the *parabolic* elements which are conjugate to $\begin{bmatrix} 1 & 1 \\ 0 & 1 \end{bmatrix}$. Since elliptic elements in Kleinian groups are torsions, our assumption implies that the groups have no elliptic elements.

Let Γ be a Kleinian group. The Lie group $PSL_2\mathbf{C}$ can be identified with the group of linear fractional automorphisms of the Riemann sphere S^2_∞, which will be regarded as the sphere at infinity of the Poincaré model of $\mathbf{H}^3$. The closure of the set of fixed points on S^2_∞ of the non-elliptic elements in Γ is called the *limit set* of Γ and denoted by Λ_Γ. The complement of Λ_Γ is called the *region of discontinuity* of Γ and denoted by Ω_Γ. The group Γ acts on Ω_Γ properly discontinuously. Ahlfors' conjecture asserts that if Γ is finitely generated, then it would be true that either $\Lambda_\Gamma = S^2_\infty$ or the Lebesgue measure of Λ_Γ is 0.

The group of linear fractional automorphisms is regarded also as the group of orientation preserving isometries of $\mathbf{H}^3$, by extending the action of the group on S^2_∞ conformally in the Poincaré model of $\mathbf{H}^3$ with S^2_∞ regarded as the points at infinity. Therefore, a Kleinian group Γ acts on $\mathbf{H}^3$ by isometries, and $\mathbf{H}^3/\Gamma$ is a hyperbolic 3-manifold if Γ is torsion free. For a hyperbolic 3-manifold $M = \mathbf{H}^3/\Gamma$, its minimal closed convex submanifold that is a deformation retract of M is called the *convex core* of M and will be denoted by C_Γ. The convex core of $\mathbf{H}^3/\Gamma$ coincides with the quotient of Nielsen's convex region, which is the convex hull of Λ_Γ in $\mathbf{H}^3$, by Γ. A finitely generated Kleinian group Γ and a hyperbolic 3-manifold $\mathbf{H}^3/\Gamma$ are said to be *geometrically finite* when the convex core C_Γ of $\mathbf{H}^3/\Gamma$ has finite volume.

By a famous lemma by Margulis, it is known that there is a positive constant ϵ_0 called the Margulis constant as follows. Let $M = \mathbf{H}^3/\Gamma$ be a hyperbolic 3-manifold, and ϵ a positive constant less than or equal to ϵ_0. Then any point $x \in M$ where the injectivity radius is smaller than $\epsilon/2$ is contained in one of the following sets which are mutually disjoint.

(1) *Margulis tube*: An open tubular neighbourhood of a closed geodesic whose length is less than ϵ.
(2) **Z**-*cusp neighbourhood*: There are a maximal parabolic subgroup P of G isomorphic to an infinite cyclic group, and an open horoball B invariant by P, which is the set of points in $\mathbf{H}^3$ translated by a generator of P at distance less than ϵ. The quotient manifold B/P, homeomorphic to $S^1 \times \mathbf{R}^2$, is embedded in M and called a **Z**-cusp neighbourhood.
(3) $\mathbf{Z} \times \mathbf{Z}$-*cusp neighbourhood*: Similarly to the precedent case, there are a maximal parabolic group P, which is isomorphic to $\mathbf{Z}\times\mathbf{Z}$ this time, and a horoball B invariant by P, which is the set of points translated by some non-trivial element of P at distance less than ϵ. The quotient manifold B/P, homeomorphic to $S^1 \times S^1 \times \mathbf{R}$, is embedded in M and called a $\mathbf{Z} \times \mathbf{Z}$-cusp neighbourhood.

The union of these subsets of M is called the ϵ-*thin part* of M, and its complement the ϵ-*thick part* of M. As we shall deal with Kleinian groups without parabolic elements, we have only to consider Margulis tubes for the thin part.

1.B. Compact cores, ends of hyperbolic 3-manifolds

For an open topological 3-manifold M, any compact 3-submanifold that is a deformation retract of M is called a *compact core* of M. Scott proved in [**42**] that for any open 3-manifold M with finitely generated fundamental group, there exists a compact core in M. It is proved in Theorem 2 of McCullough-Miller-Swarup [**30**] that a compact core of an irreducible 3-manifold M is unique in the following sense. Let C_1, C_2 be two compact cores of M, then there is a homeomorphism $f : C_1 \to C_2$ such that for the inclusions i_1, i_2 of C_1, C_2 to M, the isomorphism $(i_2)_\# \circ f_\#$ from $\pi_1(C_1)$ to $\pi_1(M)$ is equal to $(i_1)_\#$ up to inner-automorphisms of $\pi_1(M)$. We refer to this property as the *uniqueness of compact core up to homeomorphism* in this paper.

Let F be a boundary component of a compact core C. Let E be a (unique) component of the complement of C whose closure contains F. By an elementary homological argument, we can see that E has a unique end, and that E has only one frontier component, which must coincide with F. In this situation, we say that the end e *faces* the boundary component F of C. We also say that the component E faces F.

An end of hyperbolic 3-manifold is said to be *geometrically finite* when it has a neighbourhood intersecting no closed geodesics, otherwise geometrically infinite. It is easy to see that a finitely generated Kleinian group Γ is geometrically finite if and only if all the ends of $\mathbf{H}^3/\Gamma$ are geometrically finite.

For a finitely generated, torsion-free Kleinian group Γ, the quotient of the region of discontinuity Ω_Γ/Γ is a Riemann surface of finite type by Ahlfors' finiteness theorem [**1**]. The group Γ acts on $\mathbf{H}^3 \cup \Omega_\Gamma$ properly discontinuously with respect to the relative topology induced from the ordinary topology of the ball. The quotient manifold $(\mathbf{H}^3 \cup \Omega_\Gamma)/\Gamma$ is called the *Kleinian manifold* corresponding to Γ. When Γ has no parabolic elements, each component of Ω_Γ/Γ is regarded as the boundary at infinity of a geometrically finite end of $\mathbf{H}^3/\Gamma$. In particular, if $\mathbf{H}^3/\Gamma$ has no geometrically finite ends, then Ω_Γ is empty. (Refer to Morgan [**33**] for more details on these facts.)

1.C. Geodesic and measured laminations

Let S be a closed hyperbolic surface (or more generally, a complete hyperbolic surface of finite area). A closed subset of S that is a disjoint union of simple geodesics is called a *geodesic lamination.* For a geodesic lamination λ on S, a component of $S - \lambda$ is called a *complementary region* of λ. See Casson-Bleiler [**6**] for further details on geodesic laminations. We endow the space of geodesic laminations with the Hausdorff topology. In the case when S is not compact, we endow the space with the Chabauty topology, which is a generalization of the Hausdorff topology to non-compact spaces. See § 3.1 of Canary-Epstein-Green [**10**] for the precise definition of the Chabauty topology. We use the term "Chabauty topology" even for compact spaces to denote the Hausdorff topology. With this topology, the space of geodesic laminations is compact. A proof of this fact, originally due to Thurston, can be found in Proposition 4.1.7 of Canary-Epstein-Green [**10**].

A geodesic lamination λ endowed with a transverse measure t, i.e., a measure on arcs transverse to λ with the following properties is called (together with the transverse measure) a *measured lamination.*

(1) Let α, β be arcs on S both transverse to λ, an suppose that there is a homotopy $H : [0,1] \times [0,1] \to S$ such that $H(\ ,0) = \alpha, H(\ ,1) = \beta$ and each of the arcs $H(0, \ ), H(1, \ )$ is either disjoint from λ or contained in λ. Then, $t(\alpha) = t(\beta)$.
(2) For any arc α whose interior intersects λ, we have $t(\alpha) > 0$.

For a measured lamination, the geodesic lamination obtained by forgetting the transverse measure is called its *support*. (Refer to §1.7 of Penner-Harer [**6**] for more details.)

A geodesic lamination is said to be *maximal* (as a geodesic lamination) when it is not a sublamination of another geodesic lamination. This is equivalent to the condition that all the complementary regions are ideal triangles. A measured lamination is said to be *maximal* (as a measured lamination) when it is not a proper sublamination of another measured lamination. Let $\mathcal{ML}(S)$ denote the set of measured laminations on S. We should note that even when a measured lamination is maximal, its support may not be maximal as a geodesic lamination.

DEFINITION 1.1. We endow $\mathcal{ML}(S)$ with the weak topology with respect to the transverse arcs: i.e., we define $\{\lambda_k \in \mathcal{ML}(S)\}$ to converge to λ when for any arc α transverse to λ, the measure of α with respect to λ_k converges to that with respect to λ as $k \to \infty$. We call $\mathcal{ML}(S)$ with the topology defined this way the *measured lamination space* of S, and use the same symbol $\mathcal{ML}(S)$ to denote it.

The measured lamination space of S is known to be homeomorphic to the $(6g-6)$-dimensional Euclidean space, where g denotes the genus of S. This fact is first proved by Thurston [**47**], considering the notion of measured foliation equivalent to that of measure lamination. The detailed proof can be found in §8 of Fathi-Laudenbach-Poénaru [**16**]. The set of positively weighted simple closed geodesics is dense in $\mathcal{ML}(S)$, if we regard the weight as the Dirac measure supported on the simple closed geodesic. (See §6-VI of [**16**] for the proof.) By identifying two non-empty measured laminations λ_1 and λ_2 when there exists a real number $r > 0$ such that $\lambda_1 = r\lambda_2$ (where the scalar multiplication means that of the transverse measure), we obtain the *projective lamination space* $\mathcal{PL}(S)$ which is homeomorphic to the $(6g-7)$-dimensional sphere.

The definition of geodesic lamination and measured lamination depends on a hyperbolic metric on S we choose. Still, if we have two hyperbolic metrics m_1, m_2 on S, there is a natural correspondence between the geodesic laminations on (S, m_1) and those on (S, m_2), by taking a geodesic lamination on (S, m_1) to an isotopic one on (S, m_2). A proof can be found in §4.1.4 in Canary-Epstein-Green [**10**]. Similarly by isotoping also the transverse measures, we can define a correspondence between the measured laminations on (S, m_1) and those on (S, m_2). Therefore we can consider measured laminations on S without specifying a hyperbolic metric by identifying corresponding measured laminations as above. We use the term *simple closed curves* to denote simple closed geodesics regarded as a geodesic lamination to distinguish them from closed geodesics in hyperbolic 3-manifolds.

1.D. Masur domain

Next we shall define a subset of the measured lamination space called the Masur domain, which will play an important role throughout this paper, for a closed surface which is a boundary component of a compact 3-manifold. Suppose that a closed surface S of genus $g \geq 2$ is a boundary component of a compact irreducible 3-manifold N. Fix a hyperbolic structure on S, which has nothing to do with N for the moment. Let $\mathcal{C}(S)$ be the subset of $\mathcal{ML}(S)$ consisting of the weighted disjoint unions of simple closed curves on S each of which bounds a compression disc in N. Let $\overline{\mathcal{C}}(S)$ be the closure of $\mathcal{C}(S)$ with respect to the topology of $\mathcal{ML}(S)$. We define the *Masur domain* of S to be

$$\mathcal{M}(S) = \{\lambda \in \mathcal{ML}(S) | \forall \gamma \in \overline{\mathcal{C}}(S), i(\lambda, \gamma) \neq 0\}.$$

Here i denotes the *geometric intersection number*, which is defined to be

$$i(\gamma, \lambda) = \int_S \gamma \times \lambda,$$

that is, the total area of S with respect to the measure obtained by the product of the transverse measure of λ and that of γ, which becomes two-dimensional measure. In Otal [**38**], the Masur domain is defined to be a smaller set

$$\mathcal{M}'(S) = \{\lambda \in \mathcal{ML}(S) | i(\lambda, \mu) \neq 0 \text{ for } \forall \mu \text{ such that } \exists \gamma \in \overline{\mathcal{C}}(S),\ i(\mu, \gamma) = 0\},$$

when N is a boundary connected sum of two trivial I-bundles over closed orientable surfaces. We say that $\pi_1(N)$ is *uniquely freely decomposable* in this case, since the group admits an essentially unique free product decomposition. We adopt, however, the former definition of $\mathcal{M}(S)$ for all compression bodies including this special case because we need not use the properties that are valid only in $\mathcal{M}'(S)$ in this paper. Indeed, the only properties of the Masur domain that we need are those concerning pleated surfaces and freeness of the action of the modular group which will be seen next. The image of the Masur domain by the projection from $\mathcal{ML}(S)$ to $\mathcal{PL}(S)$ is called the *projectivized Masur domain* and denoted by $\mathcal{PM}(S)$. Since the projected image of $\mathcal{C}(S)$ in $\mathcal{PL}(S)$ is compact, by the definition of $\mathcal{M}(S)$, we see that both $\mathcal{M}(S)$ and $\mathcal{PM}(S)$ are open sets.

The mapping class group of S acts on $\mathcal{ML}(S)$ if we set, for an auto-diffeomorphism $f : S \to S$ and $\lambda \in \mathcal{ML}(S)$, the image $[f][\lambda]$ to be the measured lamination whose support is isotopic to $f(\lambda)$ and whose measure is equal to the transverse measure pushed-forward by f and the isotopy used to get the support. Let $\mathrm{Mod}(S, N)$ be the subgroup of the mapping class group of S consisting of the classes represented by diffeomorphisms which extend to diffeomorphisms of N. The group $\mathrm{Mod}(S, N)$ acts on $\mathcal{M}(S)$. The action is properly discontinuous except when $\pi_1(N)$ is uniquely freely decomposable. The action restricted to $\mathcal{M}'(S)$ is properly discontinuous even in the latter exceptional case. (See Otal [**38**].) We denote by $\mathrm{Mod}^0(S, N)$ the subgroup of $\text{Mod}(S, N)$ consisting of the classes whose extensions to N are homotopic to the identity as maps from N to N.

The subgroup $\text{Mod}^0(S, N)$ acts on $\mathcal{M}(S)$ freely. This is proved in Otal [**38**] unless $\pi_1(N)$ is uniquely freely decomposable. In the latter exceptional case, the group $\text{Mod}^0(S, N)$ is generated by a Dehn twist around a compression disc which is unique up to isotopy. Since every measured lamination in $\mathcal{M}(S)$ intersects the compression disc essentially, it cannot be fixed up to isotopy by the Dehn twist. Hence the action of $\text{Mod}^0(S, N)$ is free also in this case.

1.E. Pleated surfaces

Let (S, m) be a complete hyperbolic surface of finite area, and let M be a complete hyperbolic 3-manifold. A continuous map f from S to M taking cusps to cusps is said to be a *pleated surface* when there exists a geodesic lamination λ such that both $f|\lambda$ and $f|(S - \lambda)$ are locally isometric, and the path metric induced from M by f coincides with the hyperbolic metric m. The notion of pleated surface is due to Thurston (see §8 in [**45**]). A detailed explanation can be found in the chapter 5 of Canary-Epstein-Green [**10**]. Two pleated surfaces $f : (S, m_1) \to M$ and $f' : (S, m_2) \to M$ are identified as marked pleated surfaces when there exists an isometry $g : (S, m_1) \to (S, m_2)$ isotopic to the identity such that $f' = f \circ g$. We endow the space of marked pleated surfaces from S to M with a topology in such a way that $f : (S, m_1) \to M$ and $f' : (S, m_2) \to M$ are close if and only if there exists a diffeomorphism $g : S \to S$ isotopic to the identity such that $f \circ g$ and f' are close with respect to the compact-open topology and $g^* m_2$ is close to m_1 in the Teichmüller space of S. (In fact, the latter condition follows from the former condition.) When a sequence of pleated surfaces converges to a pleated surface with respect to this topology, we say that the sequence converges *as marked pleated surfaces* to make it in contrast to the geometric convergence of pleated surfaces which will appear later.

A geodesic lamination λ on S (with some hyperbolic metric) is said to be *realized* by a pleated surface $f : (S, m) \to M$ when λ is isotopic to a geodesic lamination λ' on (S, m) such that $f|\lambda'$ is totally geodesic. A measured lamination is said to be realized by a pleated surface when its support is realized. Recall that a map $f : S \to M$ is said to be *incompressible* when the homomorphism from $\pi_1(S)$ to $\pi_1(M)$ induced by f is injective. It is proved by Thurston ([**45**], see also Theorem 5.3.11 Canary-Epstein-Green [**10**]) that if M is geometrically finite and a map $f : S \to M$ is incompressible, then every measured lamination on S can be realized by some pleated surface homotopic to f. This is not the case when f is compressible. When S is a boundary component of a compact core C of a geometrically finite hyperbolic 3-manifold M, this result is generalized as follows (Corollary 2.6 in Otal [**38**]). We consider the Masur domain $\mathcal{M}(S)$ of S as a boundary component of C. If a measured lamination λ is contained in the Masur domain, there is a pleated surface homotopic to the inclusion of S which realizes λ.

Boundary components of convex cores in hyperbolic 3-manifolds are special examples of pleated surfaces. Let S be a boundary component of the convex core of a hyperbolic 3-manifold M. The surface S is bent only to one direction since it lies on the boundary of a convex set. The pleating locus of S is called the *bending locus* of S. The bending locus λ has a natural transverse measure. We define for an arc α on S transverse to λ, the measure of α to be its total curvature as an arc in M. Note that the total curvature of α is well defined even though it is not smooth. We call λ with this transverse measure *bending lamination*.

There are two essential properties of pleated surfaces which will be used in several parts of this paper; the one is the boundedness of diameters of pleated surfaces, and the other is compactness of marked pleated surfaces. These properties hold for incompressible pleated surfaces and pleated surface realizing measured laminations in the Masur domain as well. We shall here state the properties for the incompressible case.

LEMMA 1.2 (Bounded diameters). *Let $\epsilon > 0$ be a positive constant. Then there is a constant L depending only on ϵ and the topological type of S with the following property. Suppose that either $f : S \to M$ is an incompressible pleated surface. Then the diameter of the image $f(S)$ modulo the ϵ-thin part of M is bounded above by L: that is, any two points on $f(S)$ contained in the ϵ-thick part of M can be joined by an arc intersecting the ϵ-thick part of M by arcs with total length less than L.*

LEMMA 1.3 (Compactness of marked pleated surfaces). *Let K be a compact set in a complete hyperbolic 3-manifold M, which is not a surface bundle over a circle. Let $\{f_k : S \to M\}$ be a sequence of homotopic incompressible pleated surfaces. Suppose that all the images $f_k(S)$ intersect K, then $\{f_k\}$ converges uniformly on any compact set of S to a pleated surface f_∞ after taking a subsequence, which is equivalent to saying that $\{f_k\}$ converges as marked pleated surfaces to f_∞. If f_k realizes a geodesic lamination λ_k and $\{\lambda_k\}$ converges with respect to the Chabauty topology to a geodesic lamination λ_∞, then the limit surface f_∞ realizes λ_∞.*

These properties first appeared in §9 of Thurston's lecture notes [**45**]. The first lemma easily follows from the facts that the diameters of the ϵ-thick parts of hyperbolic surfaces homeomorphic to S are universally bounded and that f takes the ϵ-thin part of S into the ϵ-thin part of M since f is incompressible. A proof of the latter lemma can be found in Theorem 5.2.18 of Canary-Epstein-Green [**10**]. Otal, in §2 of [**38**], generalized these lemmata to the case of compressible surfaces. We shall state the generalizations in Chapter 4.

A geometrically infinite end e of a hyperbolic 3-manifold M facing an *incompressible* boundary component S of a compact core is said to be *simply degenerate* if there exists a sequence of pleated surfaces homotopic to S which tends to e, that is, such that for any neighbourhood U of the end e, the pleated surfaces in the sequence with sufficiently large indices are contained in U. This is equivalent to saying that there exists a sequence of simple closed curves γ_i on S such that the closed geodesic homotopic to γ_i tends to the end e as $i \to \infty$. For, on the one hand, a pleated surface containing a closed geodesic in its image can be constructed as a limit of piece-wise totally geodesic triangulated surfaces containing the closed geodesic as the image of an edge, by spinning a map around the closed geodesic infinitely many times; and on the other hand, every pleated surface can be approximated arbitrarily finely by a pleated surface realizing a closed geodesic. A detailed account of this construction can be found in pp.213-214 of Thurston [**48**]. Notice that such pleated surfaces are called ideal simplicial surfaces there.

Although the definition of a simply degenerate end can be generalized to the case when S is compressible, we reserve this term only for the case when S is incompressible throughout this paper. Bonahon proved in [**5**] that if a Kleinian group Γ without parabolic elements is freely indecomposable, i.e., admits no non-trivial free-product decomposition, then every end of $\mathbf{H}^3/\Gamma$ is either geometrically finite or simply degenerate. This implies, in particular, that $\mathbf{H}^3/\Gamma$ is almost compact. (Recall that an open 3-manifold is said to be *almost compact* if it is homeomorphic to the interior of a compact 3-manifold.)

A *simplicial ruled surface*, which is defined as follows, is an analogue of a pleated surface for a negatively curved 3-manifold. Suppose that a surface S of hyperbolic type has a triangulation T with only one vertex v, each of whose edges forms an essential loop based at the vertex. An incompressible map f from S to a negatively curved 3-manifold M is said to be a simplicial ruled surface (with respect to T) when each edge of T is mapped to a geodesic loop based at $f(v)$, each triangle is mapped to a ruled triangle, i.e., a triangle consisting of geodesic arcs with endpoints on the sides, and the cone angle at v is at least 2π. Similarly to the construction of pleated surfaces containing closed geodesics, one can construct a simplicial ruled surface containing a closed geodesic as the image of a union of an edge and a vertex. Simplicial ruled surfaces have "bounded diameters" modulo the thin part of the ambient 3-manifold just as pleated surfaces. (See Theorem 3.2.7 in Canary [**8**].)

1.F. Train tracks

A *train track* τ on a surface S is a graph consisting of C^1-edges called branches and vertices called switches satisfying the following conditions:

(1) If two branches meet at some switch, they are tangent each other there.
(2) At each switch, there exist branches in both directions.
(3) Regard complementary region (i.e., a component of the complement) of τ as a region bounded by polygons consisting of branches

and switches. Then, no complementary region is a disc with less than three corners or an annulus without corner on the boundary.

We refer readers to Harer-Penner [**40**] as a general reference for train tracks.

A regular neighbourhood N_τ of a train track τ foliated by arcs transverse to τ is called a *tied neighbourhood* of τ and the arcs forming the foliation are called *ties*. The boundary of a tied neighbourhood consists of the *horizontal boundary* which is transverse to ties and the *vertical boundary* each of whose component is a subarc of a tie. A geodesic lamination λ is said to be *carried* by a train track τ when there exists a tied neighbourhood N_τ of τ such that λ can be isotoped into N_τ so that λ becomes transverse to the ties of N_τ. When λ is a measured lamination, the transverse measure of λ determines a *weight system* on the branches of τ which satisfies the relation called the *switch condition*: that is, at each switch, the sum of the weights on the branches incoming from one direction is equal to the sum of the weights on the branches outgoing to the opposite direction. (See Penner-Harer [**40**] for further details.) Similarly another train track τ' is said to be carried by τ when it is C^1-isotoped into N_τ so that every branch transverse to the ties.

Let M be a hyperbolic 3-manifold. Following Bonahon [**5**], we say a continuous map $f : S \to M$ to be *adapted* to a tied neighbourhood N_τ of a train track τ if for each branch b of τ, its image $f(b)$ is a geodesic arc in M and f maps each tie of N_τ to a point.

For a train track τ on S with a weight system w and a map $f : S \to M$ adapted to a tied neighbourhood of τ, we define the *length* of $f(\tau, w)$ to be $\sum_{b \subset \tau} w_b \mathrm{length}(f(b))$, where b ranges over the branches of τ and w_b denotes the weight which w assigns on b. We define the *total curvature* of $f(\tau, w)$ to be $\sum_{b,b' \subset \tau} w_{b,b'} \theta(f(b), f(b'))$, where the sum is taken for the pairs of adjacent branches, $\theta(f(b), f(b'))$ denotes the exterior angle formed by the geodesic arcs $f(b), f(b')$, and $w_{b,b'}$ denotes the amount of weight which flows from b to b'. Similarly, the *quadratic variation of angle* for $f(\tau, \omega)$ is defined to be $\sum_{b,b' \subset \tau} w_{b,b'} \theta(f(b), f(b'))^2$. (Refer to §5.1 of Bonahon [**5**].)

1.G. Algebraic and geometric convergence

Let $\phi_i : \Gamma \to \Gamma_i (\subset PSL_2\mathbf{C})$ $(i = 1, 2, \ldots)$ be isomorphisms from a Kleinian group Γ. We say that $\{\phi_i\}$ converges to $\phi : \Gamma \to \Gamma'$ when it converges as representations to $PSL_2\mathbf{C}$. We also say that Γ' is an *algebraic limit* of $\{\Gamma_i\}$. Fix a Kleinian group Γ. Let (G, ϕ) and (G', ϕ') be pairs such that G, G' are Kleinian groups and $\phi : \Gamma \to G$ and $\phi' : \Gamma \to G'$ are isomorphisms as groups. The two pairs (G, ϕ) and (G', ϕ') are said to be equivalent if ϕ' is a conjugate of ϕ by an element of $PSL_2\mathbf{C}$. The set of such equivalence classes is denoted by $AH(\Gamma)$. We use the word "algebraic convergence" also for a sequence of equivalence classes of Kleinian groups in $AH(\Gamma)$ with representatives converging algebraically. The subspace of $AH(\Gamma)$ consisting of (G, ϕ) such that ϕ takes parabolic elements to parabolic elements is denoted by $AH_p(\Gamma)$.

There is another notion of convergence for Kleinian groups; *geometric convergence*. The *geometric limit* of a sequence of Kleinian groups $\{\Gamma_i\}$ is defined as follows.

DEFINITION 1.4. A Kleinian group H is called the geometric limit of $\{\Gamma_i\}$ if every element of H is the limit of a sequence $\{\gamma_i\}$ for $\gamma_i \in \Gamma_i$, and the limit of any convergent sequence $\{\gamma_{i_j} \in \Gamma_{i_j}\}$ for a subsequence $\{\Gamma_{i_j}\} \subset \{\Gamma_i\}$ is contained in H.

Suppose that for some point $x \in \mathbf{H}^3$ the minimal translation length $\inf_{\gamma \in \Gamma_i} d(x, \gamma(x))$ is bounded above independently of i. Then, $\{\Gamma_i\}$ has a geometric limit after taking a subsequence. This can be shown for instance using Gromov's compactness theorem once we see the equivalence of geometric convergence and the Gromov convergence as we shall show below. Suppose that Γ is a Kleinian group and that a sequence of faithful discrete representations $\phi_i : \Gamma \to PSL_2\mathbf{C}$ with images Γ_i converges to a faithful discrete representation with image Γ', which is the algebraic limit. Then, by Jørgensen's inequality ([**23**]), we see that $\{\Gamma_i\}$ satisfies the condition above and has a geometrically convergent subsequence. In this situation Γ' is contained in any geometric limit H of geometrically convergent subsequences of $\{\Gamma_i\}$, as we can easily see from the definition. (Refer to Jørgensen-Marden [**24**].) When $\{\Gamma_i\}$ converges both algebraically and geometrically, and the two limits coincide, we say that the sequence $\{\Gamma_i\}$ or $\{(\Gamma_i, \phi_i)\}$ *converges strongly*.

We shall explain what is the Gromov convergence using the notion of approximate isometry. (See §3.2.10 of Canary-Epstein-Green [**10**] for details.)

DEFINITION 1.5. Let (M_1, e_1) and (M_2, e_2) be two Riemannian 3-manifolds with base-frame, whose base-frames are based at $x_1 \in M_1$ and $x_2 \in M_2$ respectively. A (K, r)-*approximate isometry* from (M_1, e_1) to (M_2, e_2) is a C^∞-diffeomorphism from (X_1, x_1) to (X_2, x_2) for subsets X_1, X_2 of M_1, M_2 containing the r-balls centred at x_1, x_2 respectively such that $df(e_1) = e_2$ and

$$d_{M_1}(x, y)/K \leq d_{M_2}(f(x), f(y)) \leq K d_{M_1}(x, y) \text{ (bi-Lipschitz condition)}$$

for every $x, y \in X_1$. By perturbing approximate isometries, we can assume that both X_1 and X_2 are the r-balls themselves.

REMARK 1. The constant K in the bi-Lipschitz condition measures the distance of the differential *df* from an orthogonal map at the tangent space of every point in X_i.

Let $\{(M_i, v_i)\}$ be a sequence of complete Riemannian manifolds with base-frame. We say that $\{(M_i, v_i)\}$ *converges geometrically* (in the sense of Gromov) to a Riemannian manifold with base-frame (N, w) when for any $r > 0$ and $K > 1$ there exists an integer i_0 such that there exists a (K, r)-approximate isometry from (M_i, v_i) to (N, w) for $i \geq i_0$. This implies that there is an approximate isometry $\rho_i : X_i \to Y_i$, where X_i and Y_i contain the

r_i-balls centred at the basepoints with $r_i \to \infty$, such that $d\rho_i$ approaches an orthogonal map as $i \to \infty$. Note that if the manifolds M_i are complete hyperbolic 3-manifolds and the injectivity radii at the basepoints of M_i are uniformly bounded below by a positive constant, then the limit N is also a complete hyperbolic 3-manifold.

When $\{\Gamma_i\}$ converges geometrically to H, the hyperbolic 3-manifolds $\mathbf{H}^3/\Gamma_i$ converge geometrically to $\mathbf{H}^3/H$ in the sense of Gromov if we give base-frames which are the projections of a fixed base-frame in $\mathbf{H}^3$ by the universal coverings. Namely, if we fix a base-frame of $\mathbf{H}^3$, and define base-frames v_i of $\mathbf{H}^3/\Gamma_i$ and w of $\mathbf{H}^3/H$ which are the projections of the fixed base-frame of $\mathbf{H}^3$, then there is a (K_i, r_i)-approximate isometry from $(\mathbf{H}^3/\Gamma_i, v_i)$ to $(\mathbf{H}^3/H, w)$ where $K_i \to 1$ and $r_i \to \infty$ as $i \to \infty$.

Consider a sequence of Kleinian groups $\{\Gamma_i\}$ and pleated surfaces $f_i : (S, g_i) \to \mathbf{H}^3/\Gamma_i$ from a closed hyperbolic surface (S, g_i) with base-frame w_i set on a basepoint x_i in (S, g_i). Let y_i be $f_i(x_i)$, and v_i a base-frame on y_i. We say that $\{f_i\}$ converges geometrically to a pleated surface $f_\infty : (S', g_\infty) \to \mathbf{H}^3/G_\infty$, if

(1) (S', g_∞) is the geometric limit of (S, g_i) with base-frame w_i,
(2) the hyperbolic 3-manifold $\mathbf{H}^3/G_\infty$ is a geometric limit of $\mathbf{H}^3/G_i$ with base-frame v_i,
(3) and there exist approximate isometries $\overline{\rho}_i$ from the r_i-ball centred at x_i in (S, g_i) to (S', g_∞) and ρ_i from the r_i-ball centred at y_i in $\mathbf{H}^3/G_i$ to $\mathbf{H}^3/G_\infty$ such that $\{\rho_i \circ f_i \circ \overline{\rho}_i^{-1}\}$ converges to f_∞ uniformly on every compact subset of S'.

As is shown in Proposition 5.9 in Thurston [**48**], (see also Theorem 5.2.2 in Canary-Epstein-Green [**10**],) if there is a lower bound $\epsilon > 0$ independent of i for the injectivity radii of (S, g_i) at x_i and $\mathbf{H}^3/G_i$ at y_i, then pleated surfaces f_i as above converge geometrically after taking a subsequence. This property will be referred as the *compactness of unmarked pleated surfaces.*

A homeomorphism $\omega : S^2_\infty \to S^2_\infty$ is said to be *quasi-conformal* when ω is absolutely continuous on lines and has L^2-distributional partial derivatives $\omega_z, \omega_{\overline{z}}$ with $\|\omega_{\overline{z}}/\omega_z\|_\infty < 1$. A Kleinian group (Γ_i, ϕ_i) is said to be a *quasi-conformal deformation* of a Kleinian group Γ when Γ_i is a quasi-conformal conjugate of Γ, i.e., there exists a quasi-conformal homeomorphism $\omega_i : S^2_\infty \to S^2_\infty$ such that $\omega_i \Gamma \omega_i^{-1} = \Gamma_i$ as groups of auto-homeomorphisms acting on S^2_∞ and the isomorphism ϕ_i is induced by this conjugation. When Γ is geometrically finite, it is known that the space of quasi-conformal deformations of Γ forms an open subset of $AH_p(\Gamma)$. (Sullivan [**43**].)

Let Γ be a Kleinian group. Let $\phi : \Gamma \to \Gamma'$ be an isomorphism as abstract groups to another Kleinian group Γ'. The isomorphism ϕ induces a homotopy equivalence from $\mathbf{H}^3/\Gamma$ to $\mathbf{H}^3/\Gamma'$ since $\mathbf{H}^3$ is contractible, and therefore hyperbolic manifolds are aspherical.

CONVENTION 1. We use the corresponding capital symbol (e.g., $\Phi : \mathbf{H}^3/\Gamma \to \mathbf{H}^3/\Gamma'$) to denote a homotopy equivalence induced from an isomorphism between Kleinian groups (e.g. $\phi : \Gamma \to \Gamma'$.) Similarly we use the inverse symbol (e.g. Φ^{-1}) to denote a homotopy equivalence induced from the inverse of an isomorphism (e.g. ϕ^{-1}.)

Throughout this paper, all 3-manifolds are assumed to be orientable, even those which are not hyperbolic.

CHAPTER 2

Statements of theorems

Let us state our main theorem, which asserts that Ahlfors' conjecture is true for algebraic limits without parabolic elements of quasi-conformal deformations of a geometrically finite group without parabolic elements, and also that in this situation, Marden's conjecture is true unless the limit set is entire sphere.

THEOREM 2.1. *Let G be a finitely generated, geometrically finite Kleinian group without parabolic elements. Let $\{(G_i, \psi_i)\}$ be a sequence of quasi-conformal deformations of G with isomorphisms $\psi_i : G \to G_i$, which converges algebraically to (G', ψ) in $AH(G)$, where G' is a Kleinian group and $\psi : G \to G'$ is an isomorphism. Suppose that G' has no parabolic elements. Then the limit set $\Lambda_{G'}$ is either of Lebesgue measure 0 or the entire S^2_∞. Furthermore, $\mathbf{H}^3/G'$ is almost compact unless $\Lambda_{G'} = S^2_\infty$.*

To prove this theorem, we shall show that for such a Kleinian group G', unless the limit set is the entire sphere, each end of the hyperbolic 3-manifold $\mathbf{H}^3/G'$ has a property analogous to being simply degenerate. The property will be formulated in chap. 4 using the Masur domain. Once this is proved, we can employ an argument similar to Thurston's in [**45**] which was used to prove that geometric tameness implies the almost compactness, and prove that the manifold $\mathbf{H}^3/G'$ is almost compact. The most part of this paper is devoted to prove that if the limit set is not the entire S^2_∞, then each end of $\mathbf{H}^3/G'$ has a property analogous to being simply degenerate.

REMARK 2. Theorem 2.1 can be generalized to the case when G' has parabolic elements without changing the proof essentially if we assume that G has no accidental parabolic elements (i.e., there is no closed curve on Ω_G/G representing a parabolic element that is not homotopic to a puncture) and ψ preserves the parabolicity in both directions. To prove the main theorem in this generalized form, we need to consider instead of characteristic compression bodies which will be explained in the next section, those in a relative form, characteristic compression bodies for pared manifolds. This was done by Evans in his thesis [**13**].

REMARK 3. Otal previously announced (in a still unpublished work) that Ahlfors' conjecture is true for algebraic limits of quasi-conformal deformations of a geometrically finite group which is isomorphic to a free product of two closed surface group or an HNN-extension of a closed surface group.

CHAPTER 3

Characteristic compression bodies

We shall use the theory of characteristic compression bodies developed by Bonahon [**4**] (see also McCullough-Miller [**29**]) essentially in our proof of the main theorem. In this chapter, we shall review the theory of characteristic compression body, and prove an elementary lemma on compression bodies.

First let us recall Bonahon's theory.

A *compression body* is a possibly disconnected compact 3-manifold C with $\partial C = \partial_e C \sqcup \partial_i C$ such that no components of the *interior boundary* $\partial_i C$ are spheres and C is obtained by attaching disjoint 1-handles to a product neighbourhood of $\partial_i C$. As an exceptional case, we also regard handle bodies as compression bodies whose interior boundaries are empty. If we view C from the other side, C is constructed by attaching 2-handles and 3-handles to a product neighbourhood of $\partial_e C$ in such a way that no sphere appears as a boundary component. The *exterior boundary* $\partial_e C$ is compressible and has exactly one component in each component of C. The interior boundary $\partial_i C$ is incompressible. We do not allow the attached 1-handles to be empty, accordingly do not regard trivial I-bundles over closed surfaces as compression bodies.

Let M be an irreducible compact 3-manifold which is boundary-reducible, i.e., such that some of its boundary components is compressible. Then there exists a compression body C in M such that the exterior boundary $\partial_e C$ is the union of all the compressible components of ∂M, the interior boundary $\partial_i C$ lies in $\mathrm{Int} M$, and $M - \mathrm{Int} C$ is irreducible and boundary-irreducible. This can be easily seen by taking a maximal system of disjoint non-parallel compression discs of M and taking a regular neighbourhood of the union of the compressible boundary components and the compression discs in the system. Moreover it is proved by Bonahon that such a compression body is unique up to isotopy. We call such a compression body the *characteristic compression body* of M following Bonahon.

Now we return to the situation in our main theorem. We have a sequence of quasi-conformal deformations $\{\psi_i : G \to G_i\}$ converging to $\psi : G \to G'$. We want to get information on the ends of $\mathbf{H}^3/G'$.

Let $C(G)$ be a compact core of $\mathbf{H}^3/G$, and let $C(G')$ be that of $\mathbf{H}^3/G'$. Let F be a component of $\partial C(G)$. We shall prove in the final chapter that $\Psi|F$ is homotopic to a homeomorphism to a boundary component of the compact core $C(G')$. (Recall that $\Psi : \mathbf{H}^3/G \to \mathbf{H}^3/G'$ is a homotopy equivalence induced from the isomorphism ψ as described in Convention 1.) This can be proved more easily when F is incompressible than when F is compressible.

Assume that F is compressible. Let V be a component of the characteristic compression body containing F. Consider the covering of $\mathbf{H}^3/G$ associated to the image of $\pi_1(F)$ in $\pi_1(\mathbf{H}^3/G)$ and denote it by $\mathbf{H}^3/\Gamma$, where Γ is a Kleinian group contained in G. The surface F and the compression body V are lifted to $\mathbf{H}^3/\Gamma$ homeomorphically, and in particular the lift of V is a compact core of $\mathbf{H}^3/\Gamma$. The component of the complement of $C(G)$ facing F is also lifted homeomorphically to $\mathbf{H}^3/\Gamma$ since any closed curve in the component is homotopic to a closed curve on F. Let Γ' denote the subgroup $\psi(\Gamma)$ of G', and let Γ_i denote $\psi_i(\Gamma)$. Then we have a sequence of quasi-conformal deformations $\{\phi_i = \psi_i|\Gamma : \Gamma \to \Gamma_i\}$ converging algebraically to $\phi = \psi|\Gamma : \Gamma \to \Gamma'$. As in Convention 1, we use the symbols Φ_i, Φ to denote homotopy equivalences from $\mathbf{H}^3/\Gamma$ to $\mathbf{H}^3/\Gamma_i$ and $\mathbf{H}^3/\Gamma'$ induced from ϕ_i and ϕ above. Let C and C' be compact cores of $\mathbf{H}^3/\Gamma$ and $\mathbf{H}^3/\Gamma'$ respectively. In the middle part of our proof of the main theorem, we shall deal with a Kleinian group as Γ above such that $\mathbf{H}^3/\Gamma$ has a compact core homeomorphic to a compression body. (chapters 5-7.)

We now show a simple result on a compact 3-manifold homotopy equivalent to a compression body. An irreducible 3-manifold which is homotopy equivalent to a compression body is not always a compression body. We shall see, nonetheless, that such a 3-manifold has characteristic compression body whose complement is a union of product I-bundles over surfaces. Before stating the lemma, we should note that every compact irreducible 3-manifold that is homotopy equivalent to a connected compression body is boundary-reducible: for, it has fundamental group with non-trivial free-product decomposition.

LEMMA 3.1. *Let X be a connected compression body, and let $f : X \to Y$ be a homotopy equivalence to a compact irreducible 3-manifold Y. (As remarked above Y is then boundary-reducible.) Suppose that Y is not a compression body, and let W be a characteristic compression body of Y. Then each component of $\overline{Y - W}$ is homeomorphic to a trivial I-bundle over a closed orientable surface.*

PROOF. Suppose first, seeking a contradiction, that there is a component Z of $\overline{Y - W}$ which is not an I-bundle over a closed surface. Since X is a compression body, by applying the cut-and-paste technique to maps and compression discs of X, we can see that any incompressible map from a closed surface to X can be homotoped to a covering map into $\partial_i X$. Let $g : Y \to X$ be a homotopy inverse of f. Since ∂Z is incompressible in Y, the map $g|\partial Z$ is incompressible, hence we can homotope g so that the image of $g|\partial Z$ should be a covering map into $\partial_i X$. By Theorem 13.6 in Hempel [**22**], it follows that $g|Z$ is homotopic to a covering to X, which implies, in particular, that every component of ∂X is incompressible. This contradicts our assumption that X is a compression body. Therefore every component of $\overline{Y - W}$ must be an I-bundle over a closed surface.

Suppose next, again seeking a contradiction, that there exists a component Z of $\overline{Y - W}$ that is a twisted I-bundle over a non-orientable closed surface F, which we identify with the image of a cross section in Z. Since F is incompressible, again $g|F$ can be homotoped to a covering map onto a component T of $\partial_i X$. This is a contradiction however, because F is non-orientable and T is orientable. This completes the proof. $\square$

CHAPTER 4

The Masur domain and Ahlfors' conjecture

In this chapter, we shall generalize the notion of simple degeneracy for ends of hyperbolic 3-manifolds due to Bonahon, or geometric tameness in Thurston's term, to the case when a hyperbolic 3-manifolds has a compact core with compressible boundary. To do that, we shall make use of the Masur domain following Otal's idea to generalize the notion of pleated surfaces for such hyperbolic 3-manifolds. We shall also prove that if we generalize simple degeneracy in this way, then we can prove that the condition implies that the end has a neighbourhood homeomorphic to the product of a closed surface and an open interval, i.e., the manifold is almost compact, unless it has free fundamental group.

4.A. The main result in this chapter

The main result in this chapter is Theorem 4.1 below, which constitutes the last step of the proof of Theorem 2.1 when G is not a free group. The theorem asserts that if $\mathbf{H}^3/G'$ satisfies the condition which is a generalization of the simple degeneracy, and G' is not a free group, then both Marden's conjecture and Ahlfors' conjecture are true for the G'. In the case when G' is a purely loxodromic free group, Maskit proved in [**26**] that G' is either a Schottky group or a Kleinian group of first kind, which implies evidently that the limit set $\Lambda_{G'}$ is either of measure 0 or the entire S^2_∞. This means that Ahlfors' conjecture was already solved for such groups.

THEOREM 4.1. *Let G' be a Kleinian group without parabolic elements, which is not a free group. Let $C(G')$ be a compact core of $\mathbf{H}^3/G'$. Suppose that for each component S of $\partial C(G')$ that faces a geometrically infinite end e, there exists a sequence of simple closed curves $\{\gamma_j\}$ on S satisfying the following two conditions.*

1. *The projective classes $[\gamma_j] \in \mathcal{PL}(S)$ converge to a projective lamination $[\mu]$ contained in the projectivized Masur domain $\mathcal{PM}(S)$. We regard $\mathcal{PM}(S)$ as the entire $\mathcal{PL}(S)$ when S is incompressible.*
2. *The closed geodesic γ_j^* freely homotopic to γ_j in $\mathbf{H}^3/G'$ tends to the end e as $j \to \infty$.*

Then $\mathbf{H}^3/G'$ is almost compact. Also, the limit set $\Lambda_{G'}$ is either of Lebesgue measure 0 or the entire S^2_∞. In other words, Ahlfors' conjecture is true for such a Kleinian group G'.

We shall first prove that under the assumption of the theorem, $\mathbf{H}^3/G'$ is almost compact, i.e., homeomorphic to the interior of a compact 3-manifold (combining Proposition 4.14 with Thurston's covering theorem). Then Canary's theorem ([**7**]) implies Theorem 4.1. The method which we shall use to prove Proposition 4.14 is basically on the line of the argument in §9 of Thurston [**45**], where he proved the same fact in the case when S is incompressible. The compressibility of the boundary component of a compact core facing the end necessitates imposing the condition that the limit of the projective classes of the simple closed curves is in the projectivized Masur domain. We shall prove that the limit measured lamination μ in Theorem 4.1 is not realized by a pleated surface, and construct a sequence of pleated surfaces realizing measured laminations converging to μ. Interpolating this sequence, we shall construct a one-parameter family of negatively curved surfaces homotopic to S, which gives rise to a homeomorphism from $S \times \mathbf{R}$ to a neighbourhood of e as in Thurston's case.

We now start the proof of Theorem 4.1. As before, we consider a subgroup Γ' of G' corresponding to the image of $\pi_1(S)$ in $\pi_1(\mathbf{H}^3/G')$ by the homomorphism induced from the inclusion of S. *We assume that S is compressible* until the end of the proof of Proposition 4.14, for if it is incompressible, Proposition 4.14 below was already proved by Thurston [**45**]. (Refer also to Bonahon [**5**], Canary [**9**], and Ohshika [**35**].) Under this assumption, $\mathbf{H}^3/\Gamma'$ has a compact core C' homeomorphic to a connected compression body whose exterior boundary is S. Note that C' is not a handlebody as we assumed that G' is not a free group. The surface S and any sufficiently small neighbourhood of the end facing S are lifted homeomorphically to $\mathbf{H}^3/\Gamma'$. Then the closed geodesics γ_j^* are also lifted to closed geodesics in $\mathbf{H}^3/\Gamma'$ tending to the end that is the lift of e (i.e., the end facing the lifted S). Hence the assumptions in Theorem 4.1 are also satisfied for Γ'. We shall use the same symbols S, μ, γ_j, and γ_j^* for denoting their lifts to simplify the notations.

4.B. Realization by pleated surfaces for measured laminations on the exterior boundaries of compression bodies.

The following four lemmata, which are essentially due to the idea of Otal in [**38**], are adaptations of Thurston's argument in §9 of [**45**], where he proved that a measured lamination on the boundary of a compact core is either realizable by a pleated surface or an ending lamination, to the situation where a compact core has compressible boundary. The first lemma, Lemma 4.2, was proved in [**38**]. In Otal's paper, the smaller Masur domain $\mathcal{M}'(S)$ defined in §1.D was used for this lemma when Γ is uniquely freely decomposable. Still his proof need not this restriction. We shall explain why this is the case below.

LEMMA 4.2. *In the situation above, let $f_j : S \to \mathbf{H}^3/\Gamma'$ be a pleated surface homotopic to the inclusion, which realizes a measured lamination*

λ_j in the Masur domain $\mathcal{M}(S)$. Suppose that $\{\lambda_j\}$ converges to a measured lamination λ_∞ in $\mathcal{M}(S)$ which is maximal and connected. Suppose moreover that there exists a compact set K in $\mathbf{H}^3/\Gamma'$ which intersects all of the $f_j(S)$. Then, there are pleated surfaces f'_j equivalent to f_j such that $\{f'_j\}$ converges uniformly to a pleated surface $f_\infty : S \to \mathbf{H}^3/\Gamma'$ which realizes λ_∞.

This is Théorème 2.3 in Otal [**38**], which is a generalization of Lemma 1.3. There he further assumed that $\{\lambda_j\}$ converges in the smaller domain $\mathcal{M}'(S)$ in the case when Γ is uniquely freely decomposable. Actually, the only point where the argument for the incompressible case does not work is the proof of the boundedness theorem, a generalization of Lemma 1.2 below. The rest of the proof is the same as the incompressible case. We shall sketch the proof of Lemma 4.3 to show that the restriction to $\mathcal{M}'(S)$ is in fact unnecessary.

LEMMA 4.3 (Bounded diameters for the compressible case). *Let $\epsilon > 0$ be a constant. In the situation of Lemma 4.2, there is a constant L, which may depend on ϵ and $\{f_j\}$, such that the diameters of $f_j(S)$ modulo the ϵ-thin part of $\mathbf{H}^3/\Gamma'$ are less than L.*

PROOF. We shall sketch Otal's argument. Since the diameters of the hyperbolic surfaces homeomorphic to S modulo the two-dimensional thin part are bounded above, we have only to show that there is a positive constant ϵ' such that the ϵ'-thin part of S (with respect to the hyperbolic metric g_j induced by f_j) is taken into the ϵ-thin part of $\mathbf{H}^3/\Gamma'$. This automatically holds in the incompressible case but needs to be proved in our generalized situation.

Suppose that there is no such ϵ'. Then, after taking a subsequence, for each j, there is an annulus A_j on S contained in the $1/j$-thin part of (S, g_j) with a core curve mapped to a null-homotopic curve in $\mathbf{H}^3/\Gamma'$ by f_j. Let c_j be the simple closed curve regarded as a measured lamination homotopic to a core curve of A_j. Then its projective class $[c_j] \in \mathcal{PL}(S)$ converges to a projective lamination $[\nu]$ in $\overline{\mathcal{C}}(S)$ after passing through a subsequence. Since λ_∞ lies in $\mathcal{M}(S)$ by assumption, we have $i(\lambda_\infty, \nu) > 0$. This implies that the length of $f_j(\lambda_j)$ goes to infinity as $j \to \infty$ by the definition of the topology on the Thurston compactification of the Teichmüller space. (Refer to Exposé 8, above all Lemme II.1 of Fathi et. al. [**16**].) This is a contradiction, since the length of the measured lamination λ_j on a fixed pleated surface, say $f_1(S)$, converges to the length of λ_∞ on $f_1(S)$ as $j \to \infty$, and the length of λ_j on $f_1(S)$ is greater than or equal to that on $f_j(S)$.

Thus, we have only used the assumption that λ_∞ is in $\mathcal{M}(S)$, and it is not necessary to assume $\lambda_\infty \in \mathcal{M}'(S)$ even when Γ' is uniquely freely decomposable. □

In the case when a compact core has an incompressible boundary component S, it was proved by Thurston (refer to [**10**] and Bonahon [**5**]) that for any measured lamination λ on S, either it is realized by a pleated surface homotopic to the inclusion, or there is a sequence of pleated surfaces going to an end, which realize measured laminations converging to λ. This can be generalized to measured laminations in the Masur domain of a compressible boundary component of a compact core, which is assumed to be homeomorphic to a compression body, as follows. (The most part of this lemma was proved by Otal in [**38**].)

LEMMA 4.4. *For the exterior boundary S of a compact core C' in $\mathbf{H}^3/\Gamma'$ as above, let λ be a measured lamination in the Masur domain $\mathcal{M}(S)$. Then one of the following two alternatives holds.*

(1) *There exists a pleated surface $f : S \to \mathbf{H}^3/\Gamma'$ homotopic to the inclusion realizing λ.*
(2) *There exists a sequence of pleated surfaces $\{f_k\}$ realizing essential simple closed curves $\{c_k\}$ in $\mathcal{M}(S)$ as follows.*
    (a) *There exists a weight $w_k > 0$ on c_k such that the weighted simple closed curves $\{w_k c_k\}$ converge to λ in $\mathcal{ML}(S)$.*
    (b) *For any compact set K in $\mathbf{H}^3/\Gamma'$, the image $f_k(S)$ does not intersect K for sufficiently large k.*

PROOF. In Proposition 2.4 of Otal [**38**], it is proved that if λ in the Masur domain is not maximal and connected, there exists a pleated surface realizing λ; hence the alternative (1) above holds for such a measured lamination. (Again we need not restrict $\mathcal{M}(S)$ to $\mathcal{M}'(S)$ even in the uniquely freely decomposable case since the part where he needed to use the assumption that the lamination is in $\mathcal{M}'(S)$ is exactly Lemma 4.2 above.)

Therefore we can assume that λ is maximal and connected. Since the set of weighted essential simple closed curves is dense in $\mathcal{ML}(S)$, there exists a sequence of weighted simple closed curves $\{w_k c_k\}$ in $\mathcal{M}(S)$ which converges to λ. As we assumed that Γ' has no parabolic elements, and $w_k c_k$ is in the Masur domain, we can construct a pleated surface $\{f_k\}$ realizing c_k. By Lemma 4.2, either $\{f_k\}$ converges to a pleated surface realizing λ, or for any compact set K, the image $f_k(S)$ does not intersect K for sufficiently large k. This completes the proof. □

We shall next prove that if the alternative (2) holds in Lemma 4.4, then the pleated surfaces $f_k(S)$ tend to the end facing S.

LEMMA 4.5. *Suppose that we have a sequence of pleated surface $\{f_k\}$ homotopic to the inclusion of S such that for any compact set K, the image $f_k(S)$ is disjoint from K for sufficiently large k. Then $\{f_k\}$ tends to the end of $\mathbf{H}^3/\Gamma'$ facing S.*

PROOF. Suppose that $\{f_k\}$ as above does not tend to the end facing S. Then $\{f_k\}$ tends to another end e' of $\mathbf{H}^3/\Gamma'$ after taking a subsequence. This means, in particular, the end e' is geometrically infinite. Because the compact core C' of $\mathbf{H}^3/\Gamma'$ is a compression body and S is the only compressible boundary component, the end e' faces an incompressible boundary component S' of C'. Therefore we can apply Bonahon's argument to the end e' and see that the end e' has a neighbourhood U homeomorphic to $S' \times \mathbf{R}$, where the surface corresponding to $S' \times \{t\}$ is homotopic to the inclusion of S'. Hence for sufficiently large k, the pleated surface f_k is contained in the neighbourhood U, and homotopic to a map into S'. This implies that the inclusion of S is homotopic to a map into S' in $\mathbf{H}^3/\Gamma'$. Since C' is a compact core, such a homotopy can be taken to be contained in C'. This is impossible as the exterior boundary of a compression body cannot be homotopic into a component of the interior boundary. □

Next we shall prove that the alternatives in Lemma 4.4 are mutually exclusive.

LEMMA 4.6. *The two alternatives in Lemma 4.4 are mutually exclusive; that is, if λ can be realized by a pleated surface homotopic to the inclusion, then there is no sequence of pleated surfaces $\{f_k\}$ realizing weighted simple closed curves $w_k c_k$ which converge to λ such that for any compact set K in $\mathbf{H}^3/\Gamma'$, the image $f_k(S)$ does not intersect K for sufficiently large k.*

This means in particular that *the measured lamination μ in Theorem 4.1 cannot be realized by a pleated surface homotopic to the inclusion of S.* The proof of Lemma 4.6 will appear at the end of the next section.

4.C. Approximation by train tracks.

To prove Lemma 4.6, we need to invoke Bonahon's dichotomy for measured lamination (Proposition 5.1 in [**5**]) in terms of approximation by train tracks, which corresponds exactly the realizability/non-realizability of measured laminations in the case when S is incompressible:

PROPOSITION 4.7 (Bonahon). *Let M be a complete hyperbolic 3-manifold without cusps and S a closed hyperbolic surface. Let $f : S \to M$ be a continuous incompressible map, and λ a measured lamination on S. Then one and only one of the following two cases occurs.*

(1) *For any $\epsilon > 0$, there is a map f_ϵ homotopic to f such that* $\mathrm{length}(f_\epsilon(\lambda)) < \epsilon$.
(2) *For any $\epsilon > 0$, there are a train track τ with weight w carrying λ and a map f_ϵ homotopic to f, which is adapted to a tied neighbourhood of τ, with the following property: The total curvature and the quadratic variation of angle for $f_\epsilon(\tau, w)$ are less than ϵ. Furthermore such a map f_ϵ satisfies the following: There are $\delta > 0, t < 1$ such that $\delta \to 0, t \to 1$ as $\epsilon \to 0$, and for any simple closed curve γ such that $[\gamma]$ is sufficiently close to $[\lambda]$ in $\mathcal{PL}(S)$, the closed geodesic γ^* freely homotopic to $f(\gamma)$ in M has a part of length at least $t\mathrm{length} f_\epsilon(\gamma)$ which lies within distance δ from $f_\epsilon(\gamma)$.*

Since we cannot find an adequate literature which explicitly proves the correspondence between the alternatives realizability/non-realizability and those of Bonahon even in the incompressible case, we shall give a proof that the language of pleated surfaces can be translated into Bonahon's language even in our generalized situation.

The following lemma (Lemma 4.8) shows that if a geodesic lamination is realized by a pleated surface, then it can be approximated by a train track with geodesic branches and small exterior angles at switches such that the branches have length bounded below by a positive constant depending only on the measured lamination. The latter condition will be shown to be equivalent to one of Bonahon's alternatives.

LEMMA 4.8. *Let Γ', C', and S be a Kleinian group, a compact core of $\mathbf{H}^3/\Gamma'$, and its exterior boundary respectively defined at the end of §4.B. Suppose that a geodesic lamination ν on S is realized by a pleated surface $f : S \to \mathbf{H}^3/\Gamma'$ homotopic to the inclusion. Then there exists a positive number η depending only on f, and for any small $\epsilon > 0$, there exist a train track τ on S, which carries ν, and a continuous map $h : S \to \mathbf{H}^3/\Gamma'$ satisfying the following.*

(1) *The map h is adapted to a tied neighbourhood of τ.*
(2) *For any adjacent branches s_{n_1}, s_{n_2} of τ meeting at a switch from mutually opposite directions, the exterior angle formed by $h(s_{n_1})$ and $h(s_{n_2})$ in $\mathbf{H}^3/\Gamma'$ is less than ϵ.*
(3) *The length in $\mathbf{H}^3/\Gamma'$ of the image by h of each branch of τ is greater than η.*

PROOF. We can extend ν to a geodesic lamination ν' also realized by f, each of whose complementary region is an ideal triangle by adding extra isolated leaves if necessary. We can approximate ν' by a tied neighbourhood of a train track arbitrarily finely as follows.

First we approximate each complementary region of ν' by means of a hexagon consisting of three alternate long sides which are nearly parallel to the boundary leaves of the region, and the other three short sides which are nearly vertical to the boundary leaves. Next we delete such hexagons from S to obtain a neighbourhood of ν'. Finally we define ties in the neighbourhood as arcs nearly vertical to the leaves. Observe that the long sides of hexagons nearly parallel to the boundary leaves become the horizontal boundary of the neighbourhood, and the short sides nearly vertical to the boundary leaves become the vertical boundary. (See §1.6 of Penner-Harer [**40**] for further details.)

In this construction, for any $\delta > 0$, we can make the lengths of the ties less than δ by choosing sufficiently large hexagons. This can be shown as follows: Since we fixed a pleated surface f, the injectivity radii at the points on S with respect to the induced hyperbolic metric by f are bounded below by a positive constant. Hence by making the hexagons large as described above, and letting the area of the neighbourhood of ν go to 0, the lengths of ties can be made arbitrarily small.

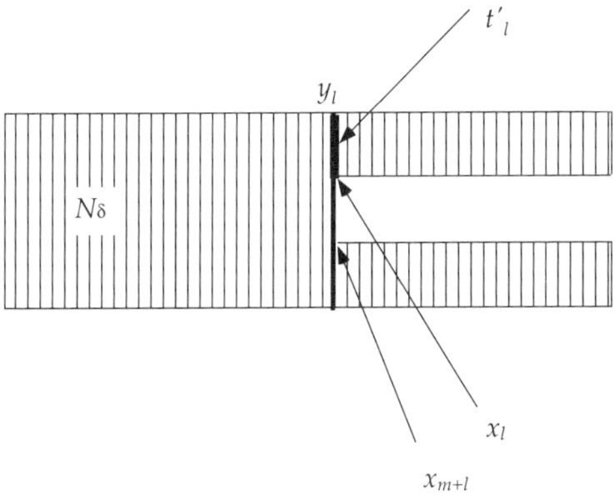

FIGURE 1

Let N_δ be a tied neighbourhood of a train track approximating ν' as above, in which the lengths of the ties are less than δ. We can assume that the train track associated with N_δ has no switches with valency more than 3 by perturbing the hexagons if necessary. By construction, the length of each component of the horizontal boundary of N_δ and the length of the part of the boundary leaf corresponding to each component of the horizontal boundary go to infinity as $\delta \to 0$, whereas the length of each component of the vertical boundary goes to 0 as $\delta \to 0$.

Let $b_1, \ldots b_m$ be the components of the horizontal boundary of N_δ. We isotope N_δ so that each b_l becomes a geodesic arc fixing its endpoints. By this isotopy, the ties can be kept not getting longer than δ, as we can see by elementary hyperbolic geometry. Let x_l, x_{m+l} denote the endpoints of b_l. Then each of them is contained in a tie, denoted by t_l or t_{m+l} respectively, which contains a component of the vertical boundary. (Recall that by definition, every vertical boundary is contained in a tie.) Let y_l be the other endpoint of the sub-tie of t_l whose interior is contained in $\mathrm{Int}N_\delta$ and whose endpoints are on ∂N_δ, one being x_l, that is, a point on t_l such that the sub-tie between x_l and y_l on t_l does not meet a vertical boundary at the interior. Let t'_l denote the sub-tie of t_l between x_l and y_l. See Figure 1.

Note that the number m is bounded by a constant depending only on the topological type of S. Also, obviously there is a positive lower bound for the lengths of b_l, which depends only on f, if we take a sufficiently small δ. We can arrange b_l and its endpoints x_l, x_{l+m}, making N_δ smaller if necessary, so that there is a positive lower bound ξ for the distance between any two of $\{x_1, \ldots, x_{2m}, y_1, \ldots, y_{2m}\}$ that correspond distinct switches of the train track associated with N_δ, since m is bounded above and the lengths of b_l are bounded below. Note that we need not make ξ smaller as δ goes to 0.

Next, we shall move endpoints of b_l to construct a train track. To make the computation of the exterior angle easier, it will be convenient to subdivide b_l so that one of the endpoints can be fixed during the move. Let $a_1, \dots a_{3m}$ be the geodesic subarcs of $b_1, \dots, b_m$ obtained by cutting $b_1, \dots b_m$ at $y_1, \dots y_{2m}$. Since we obtain rectangles foliated with the ties by cutting N_δ along the $t'_1, \dots, t'_{2m}$, it is easy to see that $\{a_1, \dots, a_{3m}\}$ is composed of $3m/2$ pairs each of which consists of two arcs parallel along ties. Renumber $a_1, \dots, a_{3m}$ so that $\{a_1, \dots, a_{3m/2}\}$ becomes a subset obtained by taking one arc from each pair. We add new points $\{z_1, \dots z_{3m/2}\}$, one on each of $\{a_1, \dots a_{3m/2}\}$, in such a way that there exists a positive lower bound η' for the distance between z_j and each of the endpoints of a_j on which z_j lies. Let $\alpha_1, \dots \alpha_{3m}$ be the subarcs of $a_1, \dots, a_{3m/2}$ obtained by cutting them at $z_1, \dots, z_{3m/2}$. Then the length of each α_j is at least η' by the choice of $z_1, \dots, z_{3m/2}$ above.

By pasting these arcs α_l together, we shall construct a train track τ_δ carried by N_δ and carrying ν', which is isotopic to the one obtained by identifying each tie of N_δ to a point if we ignore extra switches with valency 2. For that, we need to specify which arc will be moved and pasted to which arc at which point. In principle, we shall move an endpoint of each arc α_k along a tie to a point (among $y_1, \dots, y_{2m}$) on another arc, which is at the end of the tie containing the endpoint, fixing the endpoint on $z_1, \dots, z_{3m/2}$. There are two subtle problems here. First, the arc on which the target point lies must also be moved. Hence we need to specify the order of moves of arcs. Secondly, a tie on which an endpoint x_j of an arc α_k lies contains another endpoint $x_{j'}$ of another arc $\alpha_{k'}$, and we must specify which direction α_k and $\alpha_{k'}$ are to be moved; whether toward y_j or toward $y_{j'}$. To solve the first problem, we shall determine an order for the moves based on the order of subscripts. To solve the second, we shall set a rule that when two endpoints $x_j, x_{j'}$ are on the same tie and $j < j'$, then all of the $x_{j'}, y_{j'}, x_j$ are moved to y_j.

Let us describe our construction based upon these rules more formally. We denote $x_{l_1} \sim_{v_p} x_{l_2}$ when x_{l_1} and x_{l_2} are the endpoints of a component v_p of the vertical boundary of N_δ. Let $\{x_{l_1}, \dots, x_{l_m}\}$ be the subset of $\{x_1, \dots x_{2m}\}$ such that for any $x_k \in \{x_1, \dots, x_{2m}\} \setminus \{x_{l_1}, \dots, x_{l_m}\}$, there exists some x_{l_j} with $x_{l_j} \sim_{v_p} x_k$ for a component v_p of the vertical boundary and $l_j < k$. In other words, $\{x_{l_1}, \dots, x_{l_m}\}$ is obtained by choosing an endpoint of each component of the vertical boundary of N_δ that has smaller subscript than the other endpoint. We define the switches (i.e., the vertices) of τ_δ to be $\{y_{l_1}, \dots, y_{l_m}, z_1, \dots, z_{3m/2}\}$. We connect two switches y_{l_j} and z_k by a branch if and only if one of the following four conditions is satisfied. (1) Suppose that y_{l_j} and z_k are the endpoints of some α_q. Then connect y_{l_j} and z_k by a branch which is isotopic to α_q. (2) Suppose that x_{l_j} and z_k are the endpoints of some α_q. Then connect y_{l_j} and z_k by a branch which is isotopic to $\alpha_q \cup t'_{l_j}$. (3) Suppose that $x_{l_j} \sim_{v_p} x_{k'}$ and that $x_{k'}$ and z_k are

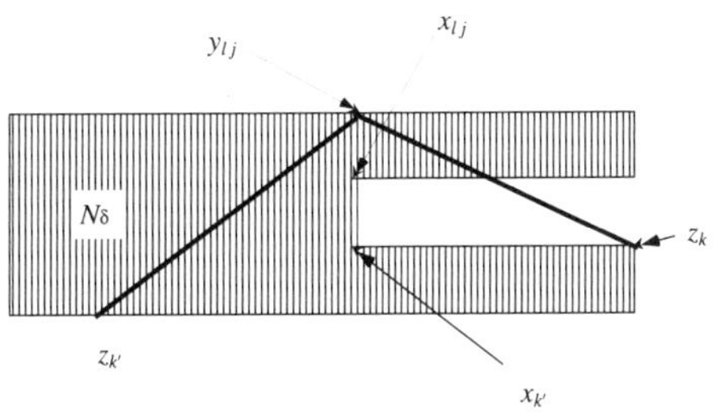

FIGURE 2

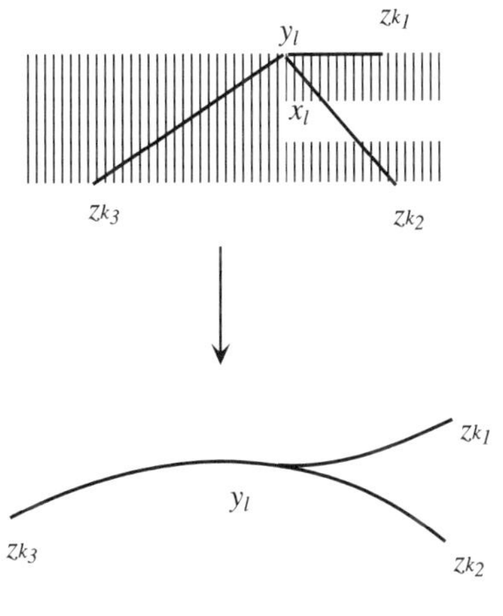

FIGURE 3

the endpoints of some α_q. Then connect y_{l_j} and z_k by a branch which is isotopic to $t'_{l_j} \cup v_p \cup \alpha_q$. (See Figure 2.) (4) Suppose that $x_{l_j} \sim_{v_p} x_{k'}$ and that $y_{k'}$ and z_k are the endpoints of some α_q. Then connect y_{l_j} and z_k by a branch which is isotopic to $t'_{l_j} \cup v_p \cup t'_{k'} \cup \alpha_q$.

To form a train track τ_δ from these branches, we define the tangency at switches naturally so that τ_δ is perturbed to be carried by the original neighbourhood N_δ. (See Figure 3.) It is easy to check that every complementary component of τ_δ is a triangle; hence τ_δ is actually a train track.

We define a piecewise geodesic map $\hat{h}$ from this train track τ_δ to $\mathbf{H}^3/\Gamma'$ as follows. Let $s_1, \ldots, s_{3m}$ be the branches of τ_δ, which are obtained by cutting τ_δ at $\{y_{l_1}, \ldots, y_{l_m}, z_1, \ldots z_{3m/2}\}$. Suppose that s_n is a branch whose endpoints are y_{l_j} and z_k. Then we define $\hat{h}(s_n)$ to be the geodesic arc in $\mathbf{H}^3/\Gamma'$ connecting $f(y_{l_j})$ and $f(z_k)$ that is homotopic to $f(s_n)$ relatively to the endpoints.

Next we shall estimate the exterior angle formed by the images of two adjacent branches of τ_δ meeting at a switch from mutually opposite directions. For that, we need to note the following facts. Let $UT(\mathbf{H}^3/\Gamma')$ be the unit tangent bundle of $\mathbf{H}^3/\Gamma'$ with the metric induced from the left invariant metric with respect to the action of $PSL_2\mathbf{C}$ on the unit tangent bundle of $\mathbf{H}^3$. First, (1) since ν' is compact and f is a pleated surface, for any $\epsilon > 0$, there exists $\delta > 0$ such that for any two points x_1, x_2 on ν' whose distance is less than δ on S, the distance in $UT(\mathbf{H}^3/\Gamma)$ between the unit tangent vectors of $f(\nu')$ at $f(x_1)$ and at $f(x_2)$ with appropriate orientations is less than $\epsilon/3$. Secondly, (2) on each complementary region of ν', the map f induces an isometric embedding from the unit tangent vector bundle $UT(S)$ of S restricted to the region to $UT(\mathbf{H}^3/\Gamma')$ because f is locally isometric on each complementary region. Thirdly, (3) if we take a sufficiently small $\delta > 0$, the geodesic arcs $b_1, \ldots, b_m$ are nearly parallel to leaves of ν'. Hence for any $z \in b_l$, and $z' \in \nu'$ which is on the boundary leaf of the complementary region containing z and within distance δ from z, the distance between the unit tangent vectors of b_l at z and of ν' at z' with appropriate orientations is less than $\epsilon/6$ in $UT(S)$. Fourthly, (4) we can easily verify, using elementary hyperbolic geometry, that for any fixed positive constant η', the following holds true. For any $\epsilon > 0$, there exists $\delta > 0$ as follows. Let $\Delta(a, b, c)$ be any geodesic triangle in $\mathbf{H}^2$ with three vertices a, b, c such that the length of the side $\overline{bc}$ is at least η' and that of the side $\overline{ab}$ is less than δ. Then both the interior angle at c and the difference between the interior angle at a and the exterior angle at b are less than $\epsilon/3$.

Now let us return to our situation, and estimate exterior angles using the facts (1)-(4) listed above. Let s_{n_1} and s_{n_2} be two adjacent branches of τ_δ which come from the mutually opposite directions. There are two possible cases: the one when the common endpoint of s_{n_1} and s_{n_2} is $y_{l_j} \in \{y_{l_1}, \ldots, y_{l_m}\}$, and the other when the common endpoint is $z_k \in \{z_1, \ldots z_{3m/2}\}$.

First assume that s_{n_1} and s_{n_2} have a common endpoint y_{l_j}. Then, the point y_{l_j} is on some $b_l \in \{b_1, \ldots, b_m\}$. We have only to consider this case under the assumption that one of the branches s_{n_1}, s_{n_2} (say s_{n_2}) is a subarc of b_l. For it is easy to see that the exterior angle in the case when neither s_{n_1} nor s_{n_2} is a subarc of b_l can be bounded by the double of the bound given under this assumption.

Let z_{k_1} and z_{k_2} be the endpoints other than y_{l_j} of s_{n_1} and s_{n_2} respectively. If z_{k_1} is on b_l, the exterior angle formed by s_{n_1} and s_{n_2} is 0. Therefore we assume that z_{k_1} is not on b_l. Then there is a $b_{l_1} \in \{b_1, \ldots b_m\} - \{b_l\}$ containing z_{k_1} one of whose endpoints, which we shall denote by w, is either

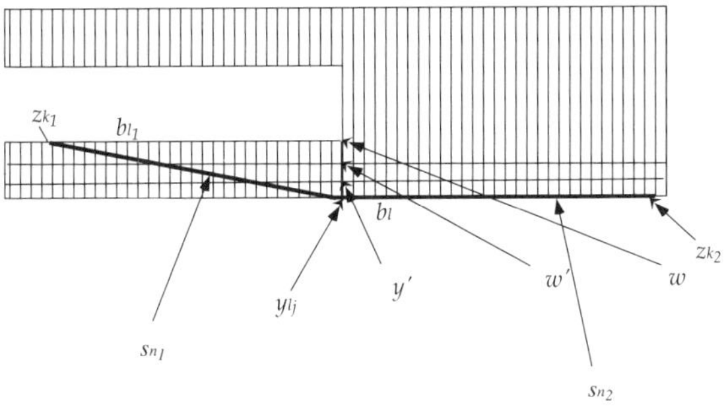

FIGURE 4

x_{l_j} or x_σ or y_σ such that $x_\sigma \sim_{v_p} x_{l_j}$ for some component v_p of the vertical boundary of N_δ in the last two cases. Since the length of each tie of N_δ is less than δ, the points w and y_{l_j} can be connected by an arc β (which is either t'_{l_j} or $t'_{l_j} \cup v_p$ or $t'_{l_j} \cup v_p \cup t'_\sigma$), whose length is bounded by 3δ. By construction, $\beta \cap \nu' \neq \emptyset$. Let y' be the point of $\beta \cap \nu'$ nearest to y_{l_j}, (hence on a boundary leaf of the complementary region containing y_{l_j},) and w' the one nearest to w. Then, if we take a sufficiently small δ, the distances of the two pairs of unit tangent vectors; $v_{y_{l_j}}$ of s_{n_2} at y_{l_j} and $v_{y'}$ of ν' at y', and v_w of b_{l_1} at w and $v_{w'}$ of ν' at w' are less than $\epsilon/6$ as was seen in (3) (if we choose appropriate orientations). Since f is locally isometric in each complementary region of ν', as was shown in (2), the distances of the two pairs of unit tangent vectors; $f_*(v_{y_{l_j}})$ of $f(b_l)$ at $f(y_{l_j})$ and $f_*(v_{y'})$ of $f(\nu')$ at $f(y')$, and $f_*(v_w)$ of $f(b_{l_1})$ at $f(w)$ and $f_*(v_{w'})$ of $f(\nu')$ at $f(w')$ are less than $\epsilon/6$. On the other hand, since the distance between w' and y' is less than δ, the distance between $f_*(v_{y'})$ and $f_*(v_{w'})$ is less than $\epsilon/3$ as was shown in (1). Thus the distance between $f_*(v_{y_{l_j}})$ and $f_*(v_w)$ is less than $2\epsilon/3$. (See Figure 4.)

Now consider the geodesic triangle Δ in $\mathbf{H}^3/\Gamma'$ formed by $\hat{h}(s_{n_1})$, the image of $b_{l_1}|[z_{k_1}, w]$ by f (whose length is at least η'), and the geodesic arc between $f(w)$ and $f(y_{l_j})$ homotopic to the union of the other two sides. Applying the fact (4) to this triangle Δ, we can see that the distance between the appropriate unit tangent vector of $\hat{h}(s_{n_1})$ at $f(y_{l_j})$ and $f_*(v_w)$ is less than $\epsilon/3$. Hence the distance between the unit tangent vector of $\hat{h}(s_{n_1})$ at $f(y_{l_j})$ and $f_*(v_{y_{l_j}})$ is less than ϵ. Since $f_*(v_{y_{l_j}})$ is a unit tangent vector of $\hat{h}(s_{n_2})$ at $f(y_{l_j})$, this implies that the exterior angle formed by $\hat{h}(s_{n_1})$ and $\hat{h}(s_{n_2})$ is less than ϵ, where $\epsilon \to 0$ as $\delta \to 0$.

Next assume that the common endpoint of s_{n_1} and s_{n_2} is $z_k \in \{z_1, \ldots, z_{3m/2}\}$. The point z_k is on some $b_l \in \{b_1, \ldots b_m\}$. Since the exterior angle formed by $\hat{h}(s_{n_1})$ and $\hat{h}(s_{n_2})$ is bounded by the sum of the exterior angle formed by $f(b_l)$ and $\hat{h}(s_{n_1})$, and that formed by $f(b_l)$ and $\hat{h}(s_{n_2})$, we have only to estimate the exterior angle formed by $f(b_l)$ and $\hat{h}(s_{n_1})$. For this, we can apply the same argument as in the case when $s_{n_1} \cap s_{n_2} = y_{l_j}$ except that we estimate the distance of the unit tangent vectors of $f(b_l)$ and of $\hat{h}(s_{n_1})$ at $f(z_k)$ using the fact (4), instead of estimating the distance of the unit tangent vector of $\hat{h}(s_{n_1})$ at $f(y_{l_j})$ and $f_*(v_w)$. Then the exterior angle is bounded by ϵ as before where $\epsilon \to 0$, as $\delta \to 0$.

Note that in the construction above, $\hat{h}(s_{n_j})$ is the geodesic arc homotopic to an arc obtained by joining two arcs; the image of α_{l_j} by f, which is a geodesic arc with the same length as that of α_{l_j} bounded below by η', and the image of an arc on a tie with length less than δ. Hence the lengths of $\hat{h}(s_{n_1})$ and $\hat{h}(s_{n_2})$ are greater than $\eta' - \delta$. We can choose δ so that $\eta' - \delta$ is positive, and set η to be $\eta' - \delta$. Then the image of each branch of τ_δ has length greater than η. By construction, we can extend $\hat{h}$ to a continuous map h from the entire surface S, which is adapted to a tied neighbourhood of τ_δ, and the proof is completed. □

The following simple lemma is fairly well known, and was in fact implicitly used in Thurston [**45**].

LEMMA 4.9. *Let S be a closed hyperbolic surface. Let $\{\lambda_k\}$ be a sequence of measured lamination on S which converges to a measured lamination λ in $\mathcal{ML}(S)$. Then the sequence of the geodesic laminations $\{|\lambda_k|\}$, which are the supports of the λ_k's, has a subsequence converging in the space of the geodesic laminations (with the Chabauty topology) to a geodesic lamination which contains the support of λ.*

PROOF. Since the geodesic lamination space with the Chabauty topology is compact, there exists a subsequence of $\{|\lambda_k|\}$ which converges to a geodesic lamination λ_∞. We denote the subsequence again by $\{|\lambda_k|\}$. We shall prove that λ_∞ contains the support of λ.

Suppose, on the contrary, that λ_∞ does not contain the support of λ. Then there exists an arc α which intersects λ transversely at the interior and does not intersect λ_∞. Since λ is a measured lamination, α has a positive measure with respect to the transverse measure of λ. By the definition of the topology of $\mathcal{ML}(S)$, the measure of α with respect to the transverse measure of λ_k converges to that with respect to the transverse measure of λ as $k \to \infty$. If follows that for sufficiently large k, the measured lamination λ_k intersects α. By the definition of the Chabauty topology, this implies that there is a point on α which is contained in the limit λ_∞. This contradicts our choice of α. □

The following lemma is a generalization of Proposition 4.7 due to Bonahon, whose expression we modified to make it convenient for our purpose.

LEMMA 4.10. *Suppose that λ is a measured lamination in $\mathcal{M}(S)$ which can be realized by a pleated surface $f : S \to \mathbf{H}^3/\Gamma'$. Let $\{w_k c_k\}$ be a sequence* *of weighted essential simple closed curves which converges to* λ. *Then for any* $\delta > 0$ *and* $t < 1$, *there exist a continuous map* $h : S \to \mathbf{H}^3/\Gamma'$ *homotopic to the inclusion and a subsequence* $\{w_{k(l)}c_{k(l)}\}$ *of* $\{w_k c_k\}$ *with the following two properties:*

(1) *The map* h *is adapted to a tied neighbourhood* N *of a train track* τ *which carries* λ *and the* $w_{k(l)}c_{k(l)}$ *for sufficiently large* l. *More strongly,* N *can be taken to contain* λ *and* $c_{k(l)}$ *in such a way that their leaves are transverse to the ties of* N.
(2) *For sufficiently large* l, *the closed geodesic* $c^*_{k(l)}$ *in* $\mathbf{H}^3/\Gamma'$ *homotopic to the image* $h(c_{k(l)})$ *has a part with length at least* $t\,\mathrm{length}\,h(c_{k(l)})$ *which is contained in the* δ*-neighbourhood of the closed curve* $h(c_{k(l)})$.

PROOF. We shall prove this lemma by using argument similar to that of Affirmation 5.12 in Bonahon [**5**], which was used to prove Proposition 4.7.

We can assume c_k to be a closed geodesic on S with a fixed hyperbolic structure. There exists a subsequence $\{c_{k(l)}\}$ of $\{c_k\}$, which converges with respect to the Chabauty topology to a geodesic lamination $\tilde{\lambda}$ which contains the support of λ by Lemma 4.9. Since λ is contained in the Masur domain, by Proposition 2.10 in Otal [**38**] stating that any geodesic lamination containing a realizable measured lamination in the Masur domain is also realizable, there exists a pleated surface f' homotopic to f which realizes $\tilde{\lambda}$. We apply Lemma 4.8 to the geodesic lamination $\tilde{\lambda}$ and the pleated surface f'. Then for any $\epsilon > 0$, a train track τ constructed in the proof of Lemma 4.8 carries $\tilde{\lambda}$ hence $c_{k(l)}$ for sufficiently large l as $\{c_{k(l)}\}$ converges to $\tilde{\lambda}$ with respect to the Chabauty topology. Let $\tau(\epsilon)$ be a train track as in Lemma 4.8 for a given $\epsilon > 0$, and $N(\epsilon)$ a tied neighbourhood as constructed there. We can isotope $\tau(\epsilon)$ and $N(\epsilon)$ so that $N(\epsilon)$ contains $\tilde{\lambda}$, hence also $c_{k(l)}$ for large l, in such a way that their leaves are transverse to the ties of $N(\epsilon)$.

The piecewise geodesic closed curve $h(c_{k(l)})$ and the closed geodesic $c^*_{k(l)}$ are freely homotopic in $\mathbf{H}^3/\Gamma'$. We can realize a homotopy by a hyperbolic simplicial annulus $A : S^1 \times I \to \mathbf{H}^3/\Gamma'$ satisfying the following conditions. (1) The vertices of A are on ∂A. (2) The images by A of the vertices on $S^1 \times \{0\}$ are the images by h of the switches of the train track $\tau(\epsilon)$ which $c_{k(l)}$ passes. (3) The valency of each vertex on $A(S^1 \times \{0\})$ is 4, that is, three faces meets at each vertex. (See the proof of Lemma 2.1 in Bonahon [**5**].)

Since A is triangulated by hyperbolic geodesic triangles and there are no vertices in the interior of $S^1 \times I$, the hyperbolic structure of $\mathbf{H}^3/\Gamma'$ induces a hyperbolic metric on $S^1 \times I$ with piecewise geodesic boundary. Note that each vertex on $S^1 \times \{0\}$ has interior angle greater than $\pi - \epsilon$ because of the estimate of the exterior angles of $h(\tau(\epsilon))$. On the other hand, the vertices on $S^1 \times \{1\}$ have interior angles at least π.

Let δ be a positive real number given in the statement of the lemma. For each point z on $S^1 \times \{0\}$ that is not a vertex, let λ_z be a geodesic arc with length δ in $S^1 \times I$, which is perpendicular to the boundary $S^1 \times \{0\}$.

If a perpendicular at z reaches $S^1 \times \{1\}$ within distance δ (with respect to the induced hyperbolic metric), the arc λ_z is defined to stop there, and its length could be less than δ. We want to specify the case when λ_z and $\lambda_{z'}$ intersect for some $z \neq z'$. Develop $S^1 \times I$ on the hyperbolic plane. Then we can see that z and z' cannot be contained in the same edge by the Gauss-Bonnet theorem. We can also see if either of the segments bounded by z and z' on $S^1 \times \{0\}$ contains an entire edge of the triangulation, which has length at least η, the arcs λ_z and $\lambda_{z'}$ cannot intersect provided that we take a sufficiently small ϵ, again by the Gauss-Bonnet theorem. Thus the remaining possibility is that z' is on an edge adjacent to the edge on which z lies. In this case, by elementary hyperbolic geometry, we can see that there exists a constant κ, which is dependent only on ϵ, δ and goes to 0 as $\epsilon \to 0$, such that if λ_z and $\lambda_{z'}$ intersect, the distance between z and z' on $S^1 \times \{0\}$ is less than κ.

Let ξ be the subset of $S^1 \times \{0\}$ consisting of the points z for which λ_z does not reach $S^1 \times \{0\}$ halfway, i.e., has length δ, and does not intersect another $\lambda_{z'}$. Then the area of $S^1 \times I$ is bounded below by the area of the points on λ_z for $z \in \xi$, which is greater than $\delta\text{length}(\xi)$. On the other hand, the area of $S^1 \times I$ is bounded above by the sum of the exterior angles which is bounded above by $\epsilon\text{length}(c_{k(l)})/\eta$ since the boundary component $S^1 \times \{1\}$ has only negative exterior angles. These imply that $\text{length}(\xi) < \epsilon\text{length}(c_{k(l)})/\eta\delta$. We can map injectively the points in $c_{k(l)} - \xi$ that are not within distance κ from the vertices, to points on $c^*_{k(l)}$ which are within the distance δ from $c_{k(l)}$, by letting z correspond to the other endpoint of λ_z. Hence the length of the points of $c^*_{k(l)}$ that are within distance δ is at least $t'\text{length}(c_{k(l)})$ where $t' = 1 - \epsilon/\eta\delta - \kappa/\eta$. As noted earlier, κ goes to 0 as $\epsilon \to 0$. Hence by taking a sufficiently small ϵ, we can make $t' \geq t$, which completes the proof of Lemma 4.10. □

PROOF OF LEMMA 4.6. The conclusion of Lemma 4.10 cannot be valid if the alternative (2) in Lemma 4.4 holds, i.e., if there exists a sequence of pleated surfaces $\{f_k\}$ realizing weighted simple closed curves $w_k c_k$, which converge to λ, such that for any compact set K in $\mathbf{H}^3/\Gamma'$, the image $f_k(S)$ does not intersect K for sufficiently large k. Lemma 4.10 tells us that in this case, λ cannot be realized by a pleated surface homotopic to the inclusion of S, hence the first alternative of Lemma 4.4 cannot hold. This shows that the two alternatives of Lemma 4.4 are mutually exclusive. □

4.D. Realization by pleated surfaces

We need two more lemmata and a corollary to prove Proposition 4.14 in the next section asserting that for our Γ' corresponding to the image of the fundamental group of S in G' in Theorem 4.1, the end of $\mathbf{H}^3/\Gamma'$ facing S has a neighbourhood homeomorphic to $S \times \mathbf{R}$, which constitutes the main part of the proof of Theorem 4.1. The first lemma below is proved in Thurston [**45**] in the special case when S is incompressible. Recall that any simple closed curve c in $\mathcal{M}(S)$ can be realized by a pleated surface, because we can construct a geodesic lamination containing c as the only compact leaf, each of whose complementary region is an ideal triangle, and a pleated surface whose pleating locus is exactly the geodesic lamination. (Refer to Figure 2.2 in Thurston [**48**] and Proposition 2.2 in Otal [**38**].)

LEMMA 4.11. *Let R be the set of measured laminations in $\mathcal{M}(S)$ which can be realized by pleated surfaces homotopic to the inclusion of S into $\mathbf{H}^3/\Gamma'$. Then R is an open dense subset of $\mathcal{M}(S)$.*

PROOF. Since this lemma can be proved by argument entirely the same as in the incompressible case, we shall only briefly sketch the proof. Since the Masur domain is an open set in $\mathcal{ML}(S)$, and the set of weighted simple closed curves is dense in $\mathcal{ML}(S)$, the weighted simple closed curves in $\mathcal{M}(S)$ are also dense. By the remark above, every weighted simple closed curve in $\mathcal{M}(S)$ lies in R; hence R is dense.

To prove that R is open, we shall show that the complement of R is closed. Suppose that $\{\lambda_j\}$ is a sequence in the complement of R which converges in $\mathcal{M}(S)$ to λ. Then by Lemma 4.4, there exists a sequence of pleated surfaces $\{f_k^j : S \to \mathbf{H}^3/\Gamma'\}$ homotopic to the inclusion that realize simple closed curves γ_k^j such that weighted simple closed curves $w_k^j\gamma_k^j$ converge to λ_j as $k \to \infty$ and such that for any compact set K in $\mathbf{H}^3/\Gamma'$, the image $f_k^j(S)$ does not intersect K for sufficiently large k. Let z' be some fixed basepoint in $\mathbf{H}^3/\Gamma'$. Then for any j, we can find a pleated surfaces $f_{k(j)}^j$ realizing simple closed curves $\gamma_{k(j)}^j$ such that the distance between $f_{k(j)}^j(S)$ and z' is greater than j, and such that for some fixed Riemannian metric on $\mathcal{ML}(S)$, the distance between $w_{k(j)}^j\gamma_{k(j)}^j$ and λ_j is less than $1/j$. Since $\{\lambda_j\}$ converges to λ, the sequence $\{w_{k(j)}^j\gamma_{k(j)}^j\}$ also converges to ν as $j \to \infty$. These imply that λ is not realizable by a pleated surface homotopic to the inclusion of S by Lemma 4.6, that is, λ is in the complement of R. Hence the complement of R is closed in $\mathcal{M}(S)$. □

The next lemma is a generalization of the compactness of the space of incompressible marked pleated surfaces, which is due to Thurston [**45**] and can be found with a detailed proof in Theorem 5.2.18 in Canary-Epstein-Green [**10**].

LEMMA 4.12. *Let ν be a measured lamination in $\mathcal{M}(S)$ all of whose complementary regions are ideal triangles and which can be realized by a pleated surface $f : S \to \mathbf{H}^3/\Gamma'$ homotopic to the inclusion of S. Let $\{w_k c_k\}$ be a sequence of weighted simple closed curves in $\mathcal{M}(S)$ which converges to ν, and let $\{f_k\}$ be a sequence of pleated surfaces realizing $\{c_k\}$. Then $\{f_i\}$ converges to f as marked pleated surfaces.*

PROOF. We have only to show that every subsequence of $\{f_k\}$ has a subsequence which converges to f. Take any subsequence $\{f_{k'}\}$ of $\{f_k\}$.

By Lemma 4.10, the images of the pleated surfaces $f_{k'}$ must intersect a fixed neighbourhood of the image of f. Then by Lemma 4.2, after taking a subsequence, the sequence $\{f_{k'}\}$ converges to a pleated surface $f' : S \to \mathbf{H}^3/\Gamma'$ which realizes the limit ν of (a subsequence of) $\{w_k c_k\}$. In this situation, what remains to prove is:

CLAIM 1. Two homotopic pleated surfaces f, f' realizing a geodesic lamination ν each of whose complementary region is an ideal triangle must be equivalent as marked pleated surfaces. (This means that there exists an auto-homeomorphism $g : S \to S$ isotopic to the identity such that $f = f' \circ g$.)

Let m be the hyperbolic structure on S induced from the map $f : S \to \mathbf{H}^3/\Gamma'$. Consider the universal covering $p : \mathbf{H}^2 \to (S, m)$, which is locally isometric. The geodesic lamination ν is covered by a geodesic lamination $\tilde{\nu}$ on $\mathbf{H}^2$ each of whose complementary regions is an ideal triangle. Let $q : \mathbf{H}^3 \to \mathbf{H}^3/\Gamma'$ be the universal covering. Then f is lifted to a map $\tilde{f} : \mathbf{H}^2 \to \mathbf{H}^3$, by which each leaf of $\tilde{\nu}$ is mapped to a geodesic. Recall that f' is homotopic to f. Therefore f' is also lifted to a map $\tilde{f}' : \mathbf{H}^2 \to \mathbf{H}^3$, which is homotopic to $\tilde{f}$ by a homotopy obtained by lifting a homotopy between f and f'. Since S is compact, we can see that for each point $x \in \mathbf{H}^2$, the distance moved by such a homotopy, i.e., the length of the trajectory, is bounded above independently of x.

Let m' be the hyperbolic structure on S induced from $\mathbf{H}^3/\Gamma'$ by f'. Since f' also realizes ν, there is a geodesic lamination ν' on (S, m') ambient isotopic to ν. This lamination ν', which may not be geodesic with respect to m, is covered by a lamination $\tilde{\nu}'$ in $\mathbf{H}^2$ by p. An ambient isotopy taking $\tilde{\nu}'$ to $\tilde{\nu}$ can be chosen to be equivariant with respect to the action of $\pi_1(S)$ and have trajectories with lengths bounded universally. These imply that each leaf ℓ' of $\tilde{\nu}'$, corresponding to a leaf ℓ of $\tilde{\nu}$, must be mapped by $\tilde{f}'$ within a bounded distance from the geodesic $\tilde{f}(\ell)$. As $\tilde{f}'(\ell')$ is also a geodesic, this means that $\tilde{f}(\ell) = \tilde{f}'(\ell')$. Moreover since each complementary region of $\tilde{\nu}$ is an ideal triangle, the image of $\mathbf{H}^2$ by $\tilde{f}$ or $\tilde{f}'$ is uniquely determined by the image of $\tilde{\nu}$ or $\tilde{\nu}'$.

The ambient isotopy taking $\tilde{\nu}'$ to $\tilde{\nu}$ gives rise to a homeomorphism $\tilde{g} : \mathbf{H}^2 \to \mathbf{H}^2$ which is equivariant with respect to the actions of $\pi_1(S)$ as covering translations. By the fact remarked above, we can assume that $\tilde{f} \circ \tilde{g} = \tilde{f}'$ by composing an equivariant isotopy preserving $\tilde{\nu}$ which can be constructed piece-by-piece on the union of complementary regions. The homeomorphism $\tilde{g}$ induces a homeomorphism $g : S \to S$ isotopic to the identity such that $f \circ g = f'$. This means that f and f' are equivalent as marked pleated surfaces. □

COROLLARY 4.13. *Let ν be a measured lamination in $\mathcal{M}(S)$ all of whose complementary regions are ideal triangles and which can be realized by a pleated surface f homotopic to the inclusion of S. Let $\{\nu_k\}$ be a sequence of measured laminations converging to ν. Then for any sufficiently large k,* *the measured lamination ν_k is realized by a pleated surface f_k homotopic to the inclusion of S and $\{f_k\}$ converges to f as marked pleated surfaces.*

PROOF. By Lemma 4.11, for any sufficiently large k, the measured lamination ν_k is realizable, hence there exists a pleated surface f_k realizing ν_k homotopic to the inclusion of S. By Lemma 4.12, for a weighted simple closed curve $w_k c_k$ sufficiently close to ν_k, we can take a pleated surface f'_k realizing c_k such that $d_{\mathbf{H}^3/\Gamma'}(f_k(x), f'_k(x)) < 1/k$ for all $x \in S$. We can take such a weighted simple closed curve $w_k c_k$ that lies within distance $1/k$ from ν_k with respect to a fixed Euclidean metric on $\mathcal{ML}(S)$. Then $\{w_k c_k\}$ converges to ν. Then again by Lemma 4.12, $\{f'_k\}$ converges to f as marked pleated surfaces. As $d(f_k(x), f'_k(x)) \leq 1/k$, the pleated surfaces $\{f_k\}$ also converge to f as marked pleated surfaces. □

4.E. A product neighbourhood of the end

The main part of the proof of Theorem 4.1, which is to show that the end of $\mathbf{H}^3/\Gamma'$ facing the lift of S has a neighbourhood homeomorphic to $S \times \mathbf{R}$, can be described as follows. The author was informed that a result analogous to this had been obtained independently by Otal in his unpublished work.

PROPOSITION 4.14. *Let G' be a Kleinian group in Theorem 4.1. Take a subgroup Γ' of G' associated to the image of the fundamental group of S such that $\mathbf{H}^3/\Gamma'$ has a compact core homeomorphic to a compression body whose exterior boundary we also denote by S. Then there exists a compact core $\tilde{C}'$ of $\mathbf{H}^3/\Gamma'$ as follows. Let E be the component of the complement of $\tilde{C}'$ that faces the boundary component homotopic to S. Then E is homeomorphic to $S \times \mathbf{R}$.*

As in §9 of Thurston [**45**], we shall first prove the following claim.

CLAIM 2. There exists a proper continuous map $G : S \times [0, 2) \to \mathbf{H}^3/\Gamma'$ such that $G(S \times \{t\})$ tends to the end facing S as $t \to 2$.

The proof of this claim, which we shall describe below, is essentially the same as that in the case when S is incompressible, whose sketch can be found in §9.4-5 in [**45**].

PROOF. Recall that we have simple closed curves γ_j in $\mathcal{M}(S)$ and closed geodesics γ_j^* in $\mathbf{H}^3/\Gamma'$ homotopic to γ_j obtained by lifting closed curves given in the assumption of Theorem 4.1. Since a neighbourhood of the end of $\mathbf{H}^3/G'$ facing S is lifted homeomorphically to a neighbourhood of the end of $\mathbf{H}^3/\Gamma'$ facing S, the closed geodesics γ_j^* tend to the end facing S. We can construct a pleated surface f_j homotopic to the inclusion of S which maps γ_j to the closed geodesic γ_j^* as noted at the beginning of §4.D. If the measured lamination μ, whose projective class is the limit of $\{[\gamma_j]\}$, is realized by a pleated surface homotopic to the inclusion of S, then by Lemma 4.10, there exists a compact set intersecting all the γ_j^*, which contradicts the fact that

$\{\gamma_j^*\}$ tends to the end facing S. This means that $\mu \in \mathcal{M}(S)$ is contained in the complement of R, where R is the set of realizable measured laminations as defined in Lemma 4.11, and in particular, that the complement of R in $\mathcal{M}(S)$ is not empty.

Since R is open and dense by Lemma 4.11, although the point μ itself may not be accessible from R by an arc, there exists a point in the complement of R in $\mathcal{M}(S)$ which is an endpoint of an arc $J : [0,1] \to \mathcal{M}(S)$ with $J[0,1) \subset R$. By Lemmata 4.4, 4.5, 4.6, we can see that a pleated surface homotopic to the inclusion of S which realizes $J(t)$ tends to the end facing S as $t \to 1$. We shall construct a proper map from $S \times [0,2)$ to a neighbourhood of the end facing S from a family of pleated surfaces realizing $\{J(t)\}$ by an "interpolation" as described in §9.5 of Thurston [**45**].

Following the idea of Thurston there, we shall define the rational depth of a measured lamination. Before the definition, we shall introduce several notions on train tracks.

DEFINITION 4.15. A train track τ is called *complete* when the following three conditions are satisfied. (This definition is due to Penner-Harer [**40**]. See)

(1) (recurrence) The train track τ supports a weight system which is positive on each branch b of τ.
(2) (transverse recurrence) For each branch b of τ, there exists an essential smooth simple closed curve σ transverse to τ which intersects b in such a way that $S - \tau - \sigma$ has no bigon component.
(3) (maximality) The τ is maximal among the train tracks satisfying the previous two conditions, i.e., it is not a proper subtrack of another train track satisfying the conditions (1) and (2).

It is easy to see that the last condition is equivalent to saying that every complementary region of τ is a triangle.

In the following argument, we need to use basic operations of train track both due to Penner-Harer, see §2.1 of [**40**].

DEFINITION 4.16. Let τ be a train track.

(1) (shifting): A homotopic move of switches and branches depicted in FIgure 5-a is called a *shifting*. (See Figure 5-a)
(2) (splitting): Let b be a branch of τ, an operation, as depicted in Figure 5-b, to split b into two and possibly adding a new branch connecting two split branches is called a *splitting* along b. There are three ways of splitting along b, called right splitting, collapsing, and left splitting respectively.

For a complete train track τ, the space V_τ of positive weight systems corresponds to an open set in $\mathcal{ML}(S)$ by taking a weight system w to a measured lamination carried by (τ, w). It can be proved that for any measured lamination, there is a complete train track carrying it with a positive weight system. (Refer to Lemma 2.6.1 and Theorem 2.8.4 in Penner-Harer

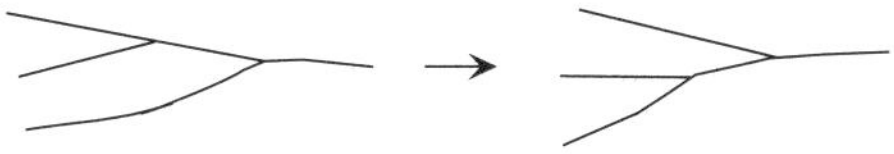

Figure 5-a

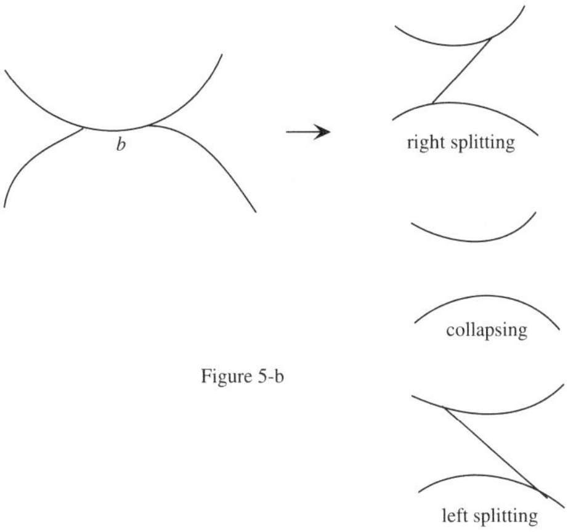

Figure 5-b

FIGURE 5

[**40**] for instance.) Moreover, we can adopt the V_τ's for complete train tracks τ as charts for $\mathcal{ML}(S)$. Since the coordinate changes between such charts are linear with rational coefficients, these charts define a PL structure on $\mathcal{ML}(S)$. The proofs of all these facts can be found in Penner-Harer [**40**].

Let τ be a complete train track carrying ν and let v_ν be a weight system on τ corresponding to ν. The system v_ν can be regarded as a point in the linear space of weight systems (allowed to have non-positive weights). Now we can define the rational depth of a measured lamination ν as follows.

DEFINITION 4.17. The *rational depth* of ν is defined to be the dimension (as a linear space over $\mathbb{Q}$) of linear functions with rational coefficients that vanish on v_ν. We call a subspace nullifying linear functions with rational coefficients a *rational subspace.*

Since the coordinate changes between charts are linear with rational coefficients, this definition is independent of the choice of a complete train track τ.

Now, we shall see the relationship between the rational depth and the type of the complementary regions of a measured lamination. The following lemma is Proposition 9.5.11 in Thurston [**45**].

LEMMA 4.18. *If a measured lamination ν on S has rational depth* 0*, then every complementary region of ν is an ideal triangle.*

PROOF. Let τ be a complete train track carrying ν. The maximality in the definition of complete train track implies that every complementary region of τ is a C^1-triangle. Suppose that ν has a complementary region R which is not an ideal triangle. Then, the boundary leaves of R are approximated by a polygonal closed curve P of branches of τ. The curve P bounds a possibly disconnected surface with C^1-boundary R' which we can isotope to be contained in R. The region R' is obtained from R by changing ideal vertices into switches and possibly squeezing it, i.e., pasting two arcs on distinct edges into one. Since R is not an ideal triangle, R' cannot be a single triangle. Therefore R' is the union of at least two complementary regions of τ'. If R' has a non-triangle component, then there is a branch b of τ' contained in R' where no leaves of ν pass. If R' has more than one triangle components, then R' was obtained by squeezing R. Hence, there is a train route, i.e., a C^1-immersed path C which has switches s_1, s_2 as the endpoints with outward branches b_1^1, b_1^2 and b_2^1, b_2^2 respectively such that all the leaves of ν coming from b_1^1 to C flows into b_2^1. (See Figure 6-a.)

The condition that the weight on b is 0 is a linear equation with rational coefficient with respect to the coordinate system associated to the rational subspace having the branches of τ as the coordinates. Since τ is recurrent, there is a measured lamination carried by non-zero weight on b. Therefore this linear equation is independent of the switch conditions. In the situation of Figure 6-a, we can deform τ by moving switches lying in the interior of C so that C becomes a single branch using the shifting operation in Definition 4.16. (See Figure 6-b.)

As is shown in Proposition 2.2.1 in [**40**], the deformed train track is still recurrent and transversely recurrent. Since obviously the maximality does not change by this operation, the deformed train track is complete. This deformation induces a coordinate change of the weight systems by a function with integer coefficients. The condition that all the leaves of ν coming from b_1^1 to C flows into b_2^1 imposes the equation $v_1^1 = v_2^1$ where $v_1^1$and v_2^1 are the b_1^1 and b_2^1-coordinates of v_ν respectively.

We shall show that this equation is linearly independent of the switch conditions. Suppose not, seeking a contradiction. This means that every weight system on τ automatically satisfies the equation $v_1^1 = v_2^1$. Let τ' be a train track obtained from τ by collapsing it along C. Since τ' is carried by τ, every measured lamination carried by τ' is also carried by τ. Then, the equation above implies that conversely every measured lamination carried by τ is also carried by τ'. On the other hand, by computing the Euler characteristic, we see that weight systems on τ' have dimension less than or equal to $-3\chi(S)-1$. Since τ is complete, its weight systems have dimension $-3\chi(S)$. This is a contradiction.

Thus, in either case, there is at least one-dimensional space of rational linear functions vanishing on v_ν. □

The next lemma is Proposition 9.5.12 of [**45**].

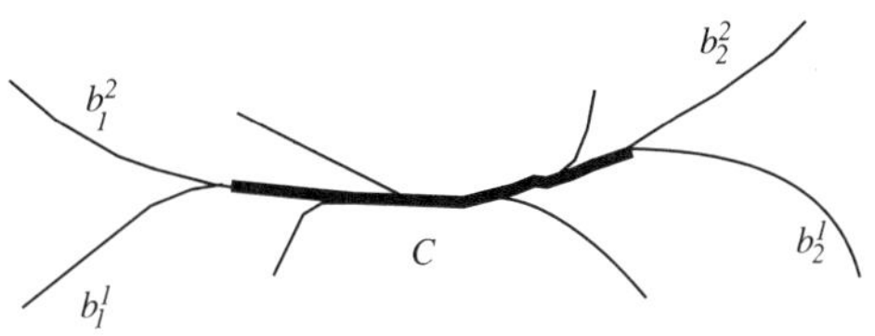

Figure 6-a

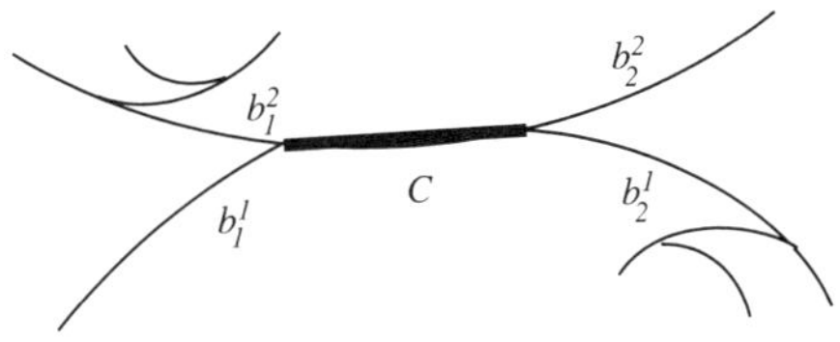

Figure 6-b

FIGURE 6

LEMMA 4.19. *If ν has rational depth 1, either every complementary region of ν is an ideal triangle or there is exactly one complementary region of ν that is not an ideal triangle, and the complementary region must be an ideal quadrilateral.*

PROOF. Suppose that ν has rational depth 1 and there is at least one complementary region that is not an ideal triangle. As in the proof of the precedent lemma, let τ be a complete train track carrying ν with weight system v_ν. Suppose, seeking a contradiction, that either ν has a complementary region R which is neither an ideal triangle nor an ideal quadrilateral, or ν has more than one complementary regions that are ideal quadrilaterals. Anyway, by the same argument as in the precedent lemma, we see that one of the following three situations occurs:

(1) There are two branches b_1, b_2 through which no leaves of ν pass.
(2) There is one branch b of τ through which no leaves of ν pass, and ν has at least one complementary region squeezed into at least two complementary regions of τ.
(3) ν has at least two complementary regions that are squeezed into at least two complementary regions of τ.

In the first situation, ν is carried by a train track τ' obtained from τ deleting the two branches b_1 and b_2. By computing the Euler number, we see the weight systems on τ' have dimension less than or equal to $-3\chi(S) - 2$. Since the weight systems of τ' allowed to have non-positive values form a linear subspace in the space of weight systems on τ, we see that v_ν is contained in a rational subspace having codimension 2. Therefore there is a two-dimensional space of rational functions vanishing on v_ν.

Next consider the second case. We can assume that β is the only one branch that does not carries a leaf of ν. As in the proof of the precedent lemma, by the shifting operation, we can assume that there is a branch C of τ, and all the leaves of ν coming from b_1^1 to C flows into b_2^1 imposes the equation $v_1^1 = v_2^1$ where $v_1^1$and v_2^1 are the b_1^1 and b_2^1-coordinates of v_ν respectively. Let τ' be a train track obtained from τ by deleting the branch η and collapsing at C. Then, the lamination ν is carried by τ'. We see that the weight systems on τ' have dimension less than or equal to $-\chi(S) - 2$. Therefore, as in the preceding case, we see that v_ν is contained in a rational space of codimension 2.

Finally consider the last case. As in the proof of the precedent lemma, by shifting τ, we can assume that there are two branches C_1, C_2 of τ; that there are branches b_1, b_1' coming to C_1 at the different endpoints, and b_2, b_2' coming to C_2 at different endpoints; and that there are equations on weights $v_1 = v_1'; v_2 = v_2'$ where v_1, v_1', v_2 and v_2' are b_1, b_1', b_2 and b_2'-coordinates of v_ν respectively. Consider then a train track τ' obtained from τ by collapsing along both C_1 and C_2. The lamination ν is carried by τ' and the weight systems on τ' have dimension less than or equal to $-\chi(S) - 2$ as before. Therefore v_ν is contained in a rational subspace of codimension 2 also in this case. □

LEMMA 4.20. *The set of measured laminations having rational depth $\geq n$ is locally a union of countably many codimension-n linear spaces.*

PROOF. Recall that the PL-structure on $\mathcal{ML}(S)$ is given by the sets of weight systems on complete train tracks. (Refer to §3.1 of Penner-Harer [**40**].) Let τ be a complete train track. Then, by definition, every measured lamination with rational depth n carried by τ lies in a codimension n rational subspace. Since there are only finitely many rational subspaces, our lemma follows. □

Therefore we can perturb the arc J to a piecewise linear path such that $J([0,1))$ contains no measured laminations with rational depth ≥ 2, and so that the measured laminations with rational depth 1 on $J([0,1])$ are countably many. Let I_0 be the set of points $t \in [0,1)$ such that every complementary region of $J(t)$ is an ideal triangle. Then for any $t \in I_0$, there exists a unique pleated surface f_t homotopic to the inclusion of S realizing $J(t)$ as in Claim 1. Moreover f_t is continuous with respect to $t \in I_0$ by Lemma 4.12.

Let t' be a point in $[0,1) - I_0$. Then by Lemma 4.18, the measured lamination $J(t')$ has positive rational depth, hence, by assumption, 1. At such a point, we need to consider and compare the left and the right limits of the pleated surfaces f_t as $t \to t' - 0$ and $t \to t' + 0$. Let us first show that both the left limit $f_{t'-0}$ and the right limit $f_{t'+0}$ exist. As the same argument applies to both the right and the left limits, we shall only consider the left limit for the time being. To show that the left limit exists, we shall prove that there is a compact set K intersecting all the f_t with $t < t'$ near t'. Suppose, on the contrary, that the images of f_t (with $t < t'$ near t') go away from every compact set. Then there exists a sequence of pleated surfaces $\{f_{t_k}\}$ tending to an end of $\mathbf{H}^3/\Gamma'$, which realize measured laminations $J(t_k)$ converging to $J(t')$ with $t_k \nearrow t'$ as $k \to \infty$. Since $J(t_k)$ is contained in the subset R of realizable laminations, by applying Lemma 4.10 to $\{f_{t_k}\}$, and by a diagonal argument, we can see that there exists a sequence of weighted simple closed curves $\{w_k\gamma_k \in \mathcal{M}(S)\}$ on S converging to $J(t')$ such that the closed geodesic γ_k^* homotopic to γ_k tends to an end as $k \to \infty$.

Let $f_k : S \to \mathbf{H}^3/\Gamma'$ be a pleated surface homotopic to the inclusion, which realizes γ_k. Then there does not exist a compact set intersecting all the images of f_k because of the following reason. If there existed such a compact set, then f_k would converge to a pleated surface after taking a subsequence by Lemma 4.2 and there would be a compact set containing all the images of pleated surfaces in a subsequence of $\{f_k\}$. This would contradict the fact that the closed geodesics γ_k^* tend to an end. Hence $\{f_k\}$ also tends to an end, and by Lemma 4.6, there is no pleated surface realizing the measured lamination $J(t')$. This is a contradiction, however, because by assumption, we have $J(t') \in R$. Thus we have proved that there exists a compact set K intersecting all the f_t for $t < t'$ near t'.

Since $J(t')$ has rational depth 1 by Lemma 4.9, there is a unique complementary region D of $J(t')$ that is not an ideal triangle, but an ideal quadrilateral, which we shall denote by Q. Let $\sigma_1, \ldots, \sigma_4$ be the four sides of the ideal quadrilateral Q in the counter-clockwise order. There are two diagonal geodesics l_1, l_2 in Q joining alternate ideal vertices. We assume that l_1 connects the ideal vertex which is an ideal endpoint of both σ_1 and σ_2, with that which is an ideal endpoint of both σ_3 and σ_4, and that l_2 is the other diagonal. A pleated surface realizing $J(t')$ maps at least one of the l_1, l_2 to a geodesic.

Since the train track τ carries $J(t')$, the ideal quadrilateral Q is also carried by τ. As we assumed that τ is complete, in particular maximal, there is no complementary region of τ that is a quadrilateral. Hence as in the proof of Lemma 4.18, in τ, the ideal quadrilateral Q is squeezed and approximated by a graph consisting of two triangles formed by three train routes, i.e. C^1-immersed paths, and one joining them. (See Figure 7.) Although the graph may not be embedded, the interiors of the complementary regions bounded by the two triangles are embedded and mutually disjoint.

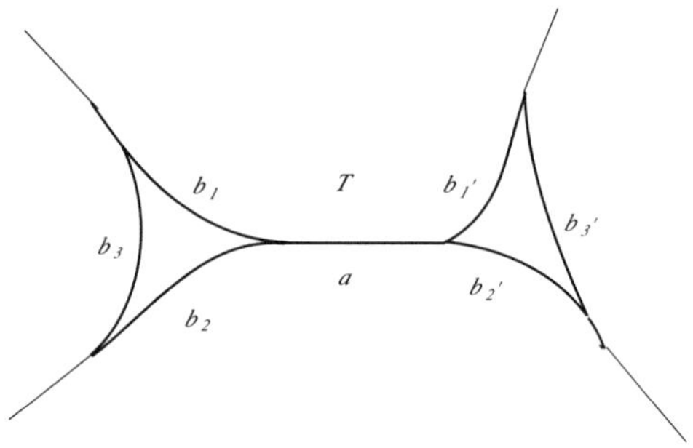

FIGURE 7

We let T denote this graph, and let a be the edge of T connecting the two triangles in the middle of T, which is a train route. There are two ways of squeezing Q; identifying subarcs of σ_1 and σ_3, or σ_2 and σ_4. Correspondingly, the edge a carries parts of either σ_1 and σ_3, or σ_2 and σ_4. We can assume, without loss of generality, that the former is the case; that is, subarcs of σ_1 and σ_3 are identified to an edge on T.

By performing the shifting operation to τ, we can assume that T is embedded and each edge of T is a branch; i.e., there are no switches on T besides six vertices as a graph, preserving the conditions that the train track τ carries $J(t')$ with positive weight system and that τ is complete. (Note that we have no branches "inside" the triangles of T since there are no bigons in the complement of τ.) We denote the branches of T other than a by $b_1, b_2, b_3; b_1', b_2', b_3'$ in such a way that the b_1, b_3, b_2 form the left triangle, (in the counter-clockwise order), and the b_1', b_2', b_3' the right one as in Figure 7. Let $w_1(t), w_2(t), w_1'(t), w_2'(t)$ be the weights on b_1, b_2, b_1', b_2' respectively for carrying $J(t)$. Then by the switch condition, an equality $w_1(t) + w_2(t) = w_1'(t) + w_2'(t)$ holds.

Since the ideal quadrilateral Q is a complementary region of $J(t')$, at the value t', we have an extra equality $w_1(t') = w_1'(t')$ which makes $J(t')$ have rational depth 1 as was shown in the proof of Lemma 4.18. Now recall that we assumed that J is piecewise linear. In particular, we can assume that J is linear with respect to the coordinates of the chart V_τ for values in $[t_0, t']$ with some $t_0 < t'$ near t'. Then the weights w_1, w_2, w_1', w_2' are linear functions of $t \in [t_0, t']$. Thus for all $t \in [t_0, t')$, we have either $w_1(t) > w_1'(t)$ or $w_1(t) = w_1'(t)$ or $w_1(t) < w_1'(t)$. Since we assumed that $J(t)$ has rational depth 0 except for countably many points, the second possibility is excluded and we have either $w_1(t) > w_1'(t)$ or $w_1(t) < w_1'(t)$. In the former case, the geometric limit of the supports of $J(t)$ as $t \to t'$ contains the diagonal l_1 of Q which corresponds to a path joining b_1, a, b_2'. In the latter case, the geometric limit contains the other diagonal l_2 corresponding to a branch path joining b_2, a, b_1'. (See Figure 8.)

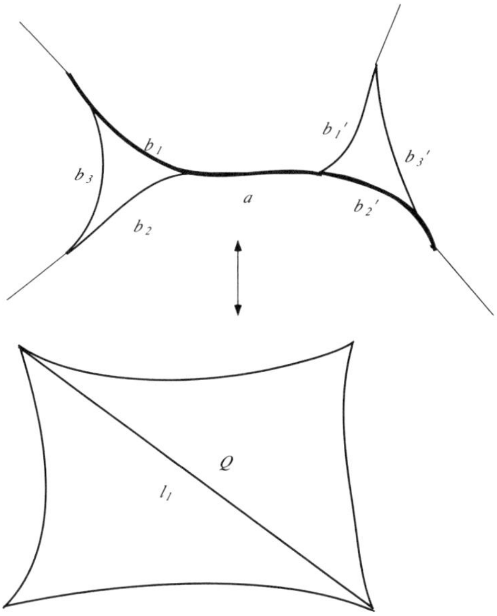

FIGURE 8

By Lemma 4.9, the left limit $\lim_{t\to t'-0}|J(t)|$ contains $|J(t')|$, the support of $J(t')$. Hence in either case, $\lim_{t\to t'-0}|J(t)|$ is uniquely determined and maximal, which is either $|J(t')|\cup l_1$ or $|J(t')|\cup l_2$. As was shown in the last paragraph, the pleated surface f_t intersects a compact set as $t\to t'-0$, and for any monotone increasing sequence $\{t_k\}$ converging to t', the pleated surfaces $\{f_{t_k}\}$ converge to a pleated surface as marked pleated surfaces after taking a subsequence. We can see that the limit pleated surface realizes the limit of the geodesic lamination $|J(t_k)|$, which must coincide with $\lim_{t\to t'-0}|J(t)|$ by Lemma 4.12. Since the limit geodesic lamination is maximal as shown above and does not depend on the choice of $\{t_k\}$ or its subsequence, the limit pleated surface is also uniquely determined independent of $\{t_k\}$ or its subsequence by Claim 1. Thus we have proved the left limit of f_t as $t\to t'-0$ exists. By exactly the same argument, we can prove the right limit of f_t as $t\to t'+0$ also exists.

If the left and the right limits coincide, then f_t is continuous at t', and there is nothing left to do. Suppose that they do not coincide. As was shown above, for each $J(t')$ with rational depth 1, there are at most two pleated surfaces $f_{t'}$ and $f'_{t'}$ that realize $J(t')$. We can assume that the left limit is $f_{t'}$ and the right limit is $f'_{t'}$. Let Q be the only complementary region of $J(t')$ that is an ideal quadrilateral as before, and l_1, l_2 the diagonal geodesics as above. We can assume that $f_{t'}$ maps l_1 to a geodesic and $f'_{t'}$ maps l_2 to a geodesic. Let D_1, D_2 be the two ideal triangles obtained by cutting Q at l_1 and D'_1, D'_2 those obtained by cutting Q at l_2. We can construct a continuous one-parameter family of simply connected surfaces with curvature ≤ -1 whose boundaries are $f_{t'}(\partial Q) = f'_{t'}(\partial Q)$ starting from $f_{t'}|Q$ and ending at $f'_{t'}|Q$ since $f_{t'}(Q)$ is formed by two ideal triangles $f_{t'}(D_1), f_{t'}(D_2)$, and $f'_{t'}(Q)$ by $f'_{t'}(D'_1), f'_{t'}(D'_2)$. (Refer to Theorem 9.5.13 of Thurston [**45**] and §3 of Canary [**9**].) This construction gives rise to a homotopy from $f_{t'}|Q$ to $f'_{t'}|Q$ relative to ∂Q. The one-parameter family of surfaces obtained by attaching this homotopy to $f_{t'}|(S - Q)$ along ∂Q is continuous since both $f_{t'}$ and $f'_{t'}$ are continuous and the image of $Q \times I$ by the homotopy lies between $f_{t'}(Q)$ and $f'_{t'}(Q)$. Thus we get a one-parameter family of surfaces $f^s_{t'}(0 \leq s \leq 1)$ with curvature ≤ -1 pleated along a geodesic lamination such that $f^0_{t'} = f_{t'}$ and $f^1_{t'} = f'_{t'}$.

For each point $t' \in [0,1) \setminus I_0$ where the right and the left limit of f_t differ as t approaches t', we interpolate a one-parameter family $\{f^s_{t'}; (s \in [0,1])\}$ of surfaces with curvature ≤ -1 as above in such a way that $f^0_{t'}$ is equal to the left limit and $f^1_{t'}$ is equal to the right limit. Thus by numbering the points of $[0,1) \setminus I_0$ where the right and the left limits differ, and setting the range of the parameter to be $1/2^n$ for the n-th point, we can make a family $f_t(t \in [0,2))$ of surfaces with curvature ≤ -1, which contains all the pleated surfaces in the original $\{f_t\}$.

It remains to prove the continuity of the interpolated family with respect to the parameter t. (Refer to §9.5, above all p. 9.50 in [**45**].) By construction, we know that the family is continuous if we restrict it to the values of parameters for which the corresponding surfaces are pleated surfaces. Also we constructed interpolations by negatively curved surfaces in such a way that in each inserted interval, the family is continuous. A problem may arise at a point in $[0,2)$ to which such inserted intervals accumulate. Let us show the continuity at such a point t_0. We have only to consider t_0 which is not in the interior of an inserted interval. We shall only show the continuity with respect to the left limit as the same argument works for the right limit. Then we can also assume that if t_0 corresponds to an inserted parameter, then t_0 is at the bottom of an inserted interval. Let $\{t_i\} \in [0,2)$ be a monotone increasing sequence which converges to t_0. The value t_0 corresponds to some original value $t'_0 \in [0,1)$. Fix a complete train track τ carrying $J(t'_0)$. As we assumed that inserted intervals accumulate to t_0, there exists a sequence $\{t'_i\}$ converging to t'_0 such that $J(t'_i)$ has depth 1. The only case that we have to consider is when we have such a sequence $\{t'_i\}$ converging to t_0 from the same side as $\{t_i\}$ does. Hence we can also assume that $\{t'_i\}$ is also monotone increasing.

The measured lamination $J(t'_0)$ has depth either 0 or 1. We consider the case when $J(t'_0)$ has depth 1 for the time being. As before, let Q be the only one complementary region of $J(t_0)$ that is an ideal quadrilateral. There is a graph T as in Figure 8 approximating Q in τ. We split the train track τ into another complete train track τ' so that the region Q is approximated by a quadrilateral $\overline{Q}$ formed by four train routes on τ', each of whose corners lie on a switch of τ'. There are two ways of splitting, right and left, depending

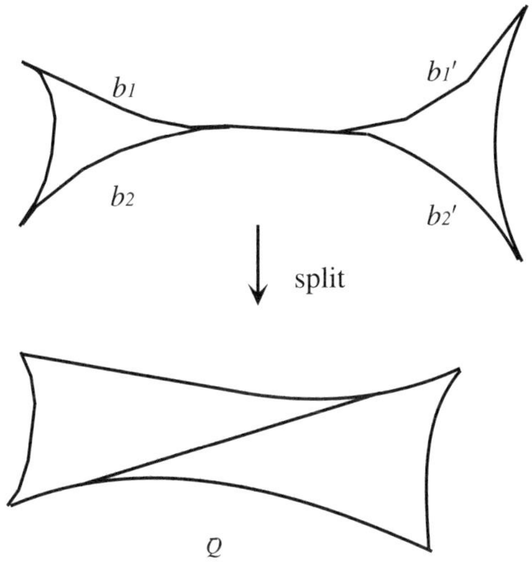

FIGURE 9

on whether τ' can carry the weight on τ with $\omega_1' > \omega_1$ or $\omega_1' < \omega_1$. As was seen before, for all t' near t_0' with $t' < t_0'$, one of the two inequalities, $\omega_1'(t') > \omega_1(t')$ or $\omega_1'(t') < \omega_1(t')$ holds independently of t'. Therefore we can choose a splitting so that τ' carries all the $J(t')$. Since t_0 was assumed to be at the bottom of an inserted interval, if $J(t_0')$ has depth 1, by our way of constructing a one-parameter family, the train track τ' also carries a maximal geodesic lamination which f_{t_0} realizes. Let ω' be the weight system on τ' corresponding to $J(t_0')$. The quadrilateral $\overline{Q}$ contains a branch inside, on which ω' vanishes. (See Figure 10.)

The measured lamination $J(t_i')$ has a unique complementary region Q_i that is an ideal quadrilateral. Since $J(t_i')$ is carried by the train track τ' for sufficiently large i, the Q_i is also carried by τ in such a way that each side is a train route and each corner lies on a switch. The quadrilateral Q_i gives rise to a rational linear relation with respect to the coordinates corresponding to τ' as in the proof of Lemma 4.18. Since J was assumed to be piece-wise linear, by taking a subsequence, we can assume that J is linear in an interval containing the t_i'. Then the linear relations caused by the $J(t_i')$ are all distinct. In particular, the quadrilaterals Q_i are all distinct, hence Q_i gets thinner and thinner as $i \to \infty$. This means that the geometric limit of the support of $J(t_i')$ is maximal; all its complementary regions are ideal triangles. Since all the measured laminations $J(t_i')$ are carried by τ', their geometric limit must also be carried by τ'. The geometric limit must contain the support of $J(t_0')$ by Lemma 4.9. Although there are two ways of extending $J(t_0')$ to a maximal geodesic lamination by adding extra leaf, the extension which is carried by τ' is uniquely determined. Since the maximal geodesic lamination realized by f_{t_0} is also carried by τ' as was seen above, it must coincide with the geometric limit of $J(t_i')$.

Recall that we have a sequence of surfaces f_{t_i} which are either pleated surface or surfaces with curvature ≤ -1. By the same argument as the proof of Lemma 4.2, we can see the surfaces f_{t_i} converge to a negatively curved surface realizing the geometric limit of $\{J(t_i')\}$. As was shown above, the geometric limit is a maximal geodesic lamination, which is uniquely determined, in the case when $J(t_0')$ has depth 1. In the case when $J(t_0')$ has depth 0, since the support of $J(t_0')$ is a maximal geodesic lamination, by Lemma 4.9, it must be equal to the geometric limit. Thus in either case, the geometric limit is a maximal geodesic lamination which is uniquely determined not depending on the choice of $\{t_i\}$ and its subsequence provided that $\{t_i\}$ converges to t_0 from below. It is easy to see a surface in the above one-parameter family f_t as we defined must be a pleated surface when it realizes a geodesic lamination all of whose complementary regions are ideal triangles. Also by Claim 1, such a pleated surface is unique up to equivalence. Hence $\{f_{t_i}\}$ must converge to f_{t_0} in either case, and we have proved the continuity of f_t with respect to the parameter when $\{t_i\}$ approaches to t_0 from below. The same argument applies to the case when $\{t_i\}$ converges to t_0 from above.

Also using the technique of taking a limit as above, we can see that for any compact set K, there exists $t_0 \in [0, 2)$ such that if $t > t_0$, the image of f_t is disjoint from K. As the original one-parameter family of pleated surfaces tends to the end facing S, and f_t is continuous with respect to t, this new one-parameter family of negatively curved surfaces $\{f_t\}$ also tends to the end facing S as $t \to 2$. Thus we have completed the proof of the claim by letting $G(x, t)$ be $f_t(x)$. □

PROOF OF PROPOSITION 4.14. We have a proper continuous map $F : S \times [1, 2) \to \mathbf{H}^3/\Gamma'$ by defining $F(x, t)$ to be $G(2t - 2)(x)$ constructed in Claim 2. We can append to this F a homotopy between f_0 and the inclusion of S to get a proper continuous map $F : S \times [0, 2) \to \mathbf{H}^3/\Gamma'$ such that $F(\ , 0) : S \to \mathbf{H}^3/\Gamma'$ is the inclusion of S. We can perturb F in such a way that it will become smooth and transverse to S fixing $F|(S \times \{0\})$. If $F^{-1}(S)$ has a component which is homeomorphic to a sphere, we can deform F and remove the component since $\pi_2(C') = \pi_2(\mathbf{H}^3/\Gamma' - C') = 0$.

In this situation, we shall prove the following claim. Let $pr : S \times [0, 2) \to [0, 2)$ be the projection to the second factor.

CLAIM 3. For any $\sigma_0 \in [0, 2)$ with $\sup pr(F^{-1}(C')) < \sigma_0$, after homotoping F only in a compact set contained in $S \times (\sigma_0, 2)$ without changing the image $F(S \times (\sigma_0, 2))$, we can construct a surface T incompressible in $\mathbf{H}^3/\Gamma' - C'$ with the following properties: Its inverse image $F^{-1}(T)$ is contained in $S \times (\sigma_0, 2)$. There is an incompressible component $\tilde{T}$ of $F^{-1}(T)$ such that $F|\tilde{T}$ is a homeomorphism onto T.

PROOF. First take $\sigma_0' \in [0,2)$ so that $F(S \times [\sigma_0', 2))$ is disjoint from $F(S \times [0, \sigma_0])$. Such σ_0' exists because of the properness of F. We also take $\sigma_0'' \in [\sigma_0', 2)$ and $\tau_0 \in (\sigma_0', \sigma_0'')$. We have a (possibly) singular surface $F(S \times \{\tau_0\})$ in $F(S \times [\sigma_0', \sigma_0''])$. We can assume that $F(S \times [\sigma_0', \sigma_0''])$ is a codimension-0 submanifold of $\mathbf{H}^3/\Gamma'$. Then the second homology class of $F(S \times [\sigma_0', \sigma_0''])$ represented by $F(S \times \{\tau_0\})$, which is non-trivial because S represents a non-trivial second homology class in C' as we assumed that C' is not a handlebody, has singular Thurston norm less than or equal to $|\chi(S)|$.

Since the singular Thurston norm is equal to the Thurston norm as was proved in Corollary 6.18 of Gabai [**19**], there is an embedded surface $\overline{T}$ homologous to $F(S \times \{\tau_0\})$ in $F(S \times [\sigma_0', \sigma_0''])$ with Thurston norm less than or equal to that of $|\chi(S)|$. Since $\overline{T}$ is homologous to S in $\mathbf{H}^3/\Gamma'$, it separates the end of $\mathbf{H}^3/\Gamma'$ facing S from C'. There is no non-separating surface in $\mathbf{H}^3/\Gamma' - C'$: For if there were, there would be a closed curve intersecting such a surface at one point, which would contradicts the homotopy invariance of the algebraic intersection number, since the closed curve would be homotopic into C' to become disjoint from the surface. Therefore, at least one component of $\overline{T}$, which we denote by T, separates the end facing S from C'. Note that the genus of T is less than or equal to that of S since it is a component of $\overline{T}$ having Thurston norm less than or equal to $|\chi(S)|$. We perturb F to make it transverse to T. Then $F^{-1}(T)$ is a surface embedded in $S \times (\sigma_0, 2)$ by our choice of σ_0'. There must be a component $\tilde{T}$ of $F^{-1}(T)$ which separates $S \times \{0\}$ from the end $S \times \{2\}$ since otherwise, there would be an infinite arc starting at a point on S tending to the end facing S which does not intersect T, and it would contradict the fact that T separates the end from C'. Since T represents a non-trivial second homology class (for it separates the end facing S from C'), $F|\tilde{T}$ has non-zero degree.

Since C' is a compact core, we can homotope T in $\mathbf{H}^3/\Gamma'$ to a smooth map into C', which we denote by $j : T \to C'$. Consider the projection of $\tilde{T}$ to $S \times \{0\}$ by the canonical projection of $S \times [0,2)$ to $S \times \{0\}$. Because $\tilde{T}$ is separating, this projection has non-zero degree, hence induces a surjection from $\pi_1(\tilde{T})$ to $\pi_1(S \times \{0\})$. Since $F|(S \times \{0\}) : S \times \{0\} \to \mathbf{H}^3/\Gamma$ is a homeomorphism to S and the fundamental group of S surjects to that of $\mathbf{H}^3/\Gamma'$, we can see that $(F|\tilde{T})_\# : \pi_1(\tilde{T}) \to \pi_1(\mathbf{H}^3/\Gamma')$ is surjective. Since $F|\tilde{T}$ induces a homomorphism from $\pi_1(\tilde{T})$ to $\pi_1(T)$ whose composition with $j_\#$ is conjugate to $(F|\tilde{T})_\# : \pi_1(\tilde{T}) \to \pi_1(\mathbf{H}^3/\Gamma')$, this means that $j_\#$ must also be surjective. We shall next prove that T has genus greater than or equal to that of S using the fact that $j_\#$ is surjective.

Consider a system of disjoint, non-parallel, compressing discs $\{D_1, \dots, D_k\}$ which decomposes C' into product I-bundles over closed surfaces. Make j transverse to the discs of the system, and consider the inverse image $J = j^{-1}(D_1) \sqcup \dots \sqcup j^{-1}(D_k)$, all of whose components we can assume to be essential on T. We perform surgery on T along the simple closed curves of J, that is, we cut T along each component of J and attach a disc to every boundary component of the resulting surface. Let $\overline{T}_1, \ldots, \overline{T}_l$ be the components of the surface that we get by the surgery. Note that every component of J is mapped to a null-homotopic closed curve in C' by j. Therefore j induces a smooth map $j' : \overline{T}_1 \sqcup \cdots \sqcup \overline{T}_l \to C'$ such that for each $\overline{T}_\lambda$ $(\lambda = 1, \ldots, l)$, the restriction $j'|\overline{T}_\lambda$ is disjoint from the discs $D_1, \ldots D_k$, hence is homotopic to a map to an interior boundary component of C'. Note that the fundamental group of C' contains, as a factor of a free product decomposition, the free product of the fundamental groups of the interior boundary components. Since $j_\#$ is surjective, for each interior boundary component Σ of C', there is some $\overline{T}_\lambda$ such that $j'|\overline{T}_\lambda$ is homotopic to a non-zero degree map to Σ. For such λ, the genus of $\overline{T}_\lambda$ must be greater than or equal to that of Σ.

On the other hand, each non-separating disc among $D_1, \ldots, D_k$ is a co-core disc of a 1-handle attached to $\partial_i C' \times \{1\}$ to form C' as a compression body, which corresponds to an infinite cyclic factor of a free product decomposition of $\pi_1(C')$. Therefore if some of such discs are disjoint from T, then $j_\#$ cannot be surjective. Then by using ordinary technique of handle addition, or considering the 1-dimensional homology group, we can see that each non-separating disc contributes at least 1 to the genus of T. Hence in this situation, the genus of T is at least the sum of the genera of incompressible boundary components of C' plus the number of the non-separating discs among $D_1, \ldots, D_k$, which is equal to the genus of S. Thus we have proved that T has genus greater than or equal to that of S. Since we already know that the genus of T is less than or equal to that of S, we see that the genera of S and T are equal, i.e., they are homeomorphic.

Next we shall show that T is incompressible in $\mathbf{H}^3/\Gamma' - C'$. Let W be the codimension-0 submanifold cobounded by $\partial_i C'$ and T. Suppose, seeking a contradiction, that T is compressible in $\mathbf{H}^3/\Gamma' - C'$. Then there is a compressing disc D for T disjoint from C'. First we consider the case when D lies outside W. By performing the surgery along D, we get a submanifold W' containing W with a boundary component $T_{W'}$ which separates C' from the end facing T, one of the components of the surface obtained from T by compression. The genus of $T_{W'}$ is less than that of T. Let E' be the component of the complement of W' facing $T_{W'}$. Since $F(S \times \{t\})$ is contained in E' for t sufficiently close to 2, we see that the fundamental group of E' surjects to that of $\mathbf{H}^3/\Gamma'$. Since every loop in E' (with basepoint on $T_{W'}$) is homotopic to a loop in C' (conjugated by a fixed arc connecting the basepoint on $T_{W'}$ and that in C'), we see that the fundamental group of $T_{W'}$ also surjects to that of $\mathbf{H}^3/\Gamma'$. Then by the argument which we applied for T above, we see that $T_{W'}$ has genus greater than or equal to that of S, hence that of T. This is a contradiction.

Next, if D is a non-separating compressing disc contained in W, then there is a simple closed curve on T intersecting D at one point, which contradicts the homotopy invariance of the algebraic intersection number since the closed curve is homotoped into C' to become disjoint from D.

Finally suppose that D is a separating compressing disc in W. Then, W is decomposed into the boundary connected sum of two submanifolds W_1 and W_2, one of which, say W_1, contains C'. The boundary of W_1, which we denote by T_{W_1} has genus less than that of T. The fundamental group of T_{W_1} surjects to that of $\mathbf{H}^3/\Gamma'$ as we can see by the same argument for $T_{W'}$ since W_1 contains C', and it follows that the genus of T_{W_1} is greater than or equal to that of S by the same argument as for T. This is a contradiction.

It remains to show that we can arrange so that $\tilde{T}$ is incompressible and $F|\tilde{T}$ is a homeomorphism. If $\tilde{T}$ is incompressible, then it is homeomorphic to S since it is embedded in $S \times [0, 2)$. Since $F|\tilde{T}$ has a non-zero degree map between two homeomorphic surfaces, we can homotope F along $\tilde{T}$ keeping its image intact so that $F|\tilde{T}$ is a homeomorphism. Thus, we have only to show that $\tilde{T}$ can be arranged to be incompressible. We should first recall that by our choice of σ_0', the surface $\tilde{T}$ is contained in $S \times (\sigma_0, 2)$. Suppose that $\tilde{T}$ is compressible. Then there is a compressing disc D which is contained in $S \times (\sigma_0, 2)$. Consider the singular disc $F(D)$, whose boundary is contained in T. If $F(D)$ is relatively inessential in $(\mathbf{H}^3/\Gamma', T)$, i.e., the disc can be homotoped into T fixing the boundary, then we can easily deform F in a neighbourhood of D (without changing the image $F([\sigma_0, 2))$ so that $F^{-1}(T)$ contains a surface obtained by compressing $\tilde{T}$ along D. We take a component of the resulting surface which separates $S \times \{0\}$ from $S \times \{2\}$. If we obtain an incompressible surface by repeating this deformation, then we are done.

Suppose that we have performed compression repeatedly until there are no more such compressing discs mapped to inessential discs in $(\mathbf{H}^3/\Gamma', T)$, and that still $\tilde{T}$ is compressible. Then there is a compressing disc D for $\tilde{T}$ contained in $S \times (\sigma_0, 2)$ which is mapped to a singular disc that is essential relatively in $(\mathbf{H}^3/\Gamma', T)$. Note that the singular disc $F(D)$ does not intersect C' because D is contained in $S \times (\sigma_0, 2)$, and σ_0 is greater than $\sup prF^{-1}(C')$. By the analytic loop theorem, which was proved by Meeks-Yau [**31**], we can see that there exists a compressing disc D' for T which is contained in an arbitrarily small neighbourhood N_D of $F(D)$. This means that T is compressed outside C', and we have reached a contradiction. Thus we can make $\tilde{T}$ incompressible and have completed the proof. □

REMARK 4. In an older version of the manuscript of this paper, there was a gap in the proof of the claim above. The argument invoking Gabai's work on the singular Thurston norm was added later during the process of revising the paper to fill this gap. Prior to this revision, Souto used the same technique involving the Thurston norm in his paper [**44**] in a similar line of argument to construct a product structure near an end. The author thanks the referee for drawing his attention to Souto's work.

Since $\tilde{T}$ is incompressible and embedded, it is isotopic to $S \times \{0\}$, and by the claim above it is contained in $S \times (\sigma_0, 2)$ which is disjoint from $F^{-1}(C')$. Let B' be the codimension-0 submanifold of $\mathbf{H}^3/\Gamma'$ bounded by $S \sqcup T'$. Let $\tilde{C}'$ be $C' \cup B'$, and take $\sigma_1 \in [0, 2)$ greater than both $(\sigma_0 + 2)/2$ and $\sup pr F^{-1}(\tilde{C}')$. Then by applying the claim above replacing C' by $\tilde{C}'$ and σ_0 by σ_1, we get a surface T_1 embedded in $\mathbf{H}^3/\Gamma' - \tilde{C}'$ as an incompressible surface such that there is an incompressible component $\tilde{T}_1$ of $F^{-1}(T_1)$ contained in $S \times (\sigma_1, 2)$ and $F|\tilde{T}_1$ is a homeomorphism onto T_1. Note that in the proof the claim, the property that C' is a compact core is used only in a restricted form that the fundamental group of C' surjects to that of $\mathbf{H}^3/\Gamma'$. Therefore $\tilde{C}'$ can play the role of C' although we have not yet proved that $\tilde{C}'$ is a compact core.

Note that $\tilde{T}$ and $\tilde{T}_1$ are isotopic. Furthermore, an isotopy between them can be taken to be contained in $S \times (\sigma_0, 2)$, hence to be disjoint from $F^{-1}(C')$. Thus T and T_1 are disjoint surfaces which are homotopic in $\mathbf{H}^3/\Gamma' - C'$ and incompressible in $\mathbf{H}^3/\Gamma' - C'$. This implies that a homotopy between them can be deformed so that its image is contained in a submanifold P_1 cobounded by T and T_1. It follows furthermore that P_1 is homeomorphic to $S \times [0, 1]$ by using classical three-dimensional topology. Next we take a number σ_2 greater than both $(\sigma_1 + 2)/2$ and $\sup pr F^{-1}(\tilde{C}' \cup P_1)$, and repeat the argument replacing T with T_1, $\tilde{T}$ with $\tilde{T}_1$, and $\tilde{C}'$ with $\tilde{C}' \cup P_1$. Then we get a surface T_2 contained in $F(S \times (\sigma_2, 2))$ which is incompressible in $\mathbf{H}^3/\Gamma' - (\tilde{C}' \cup P_1)$ and cobounds a 3-submanifold homeomorphic to $S \times [1, 3/2]$ with T_1. Repeating this procedure, we get at the k-th step an embedding H_k from $S \times [0, 2 - 1/2^{k-1}]$ into $\mathbf{H}^3/\Gamma' - C'$ such that $S \times \{2 - 1/2^{k-1}\}$ corresponds to a surface T_k lying outside $F(S \times [0, \sigma_k])$ which is incompressible in $\mathbf{H}^3/\Gamma' - C'$. Thus, we get an embedding from $S \times [0, 2)$. Since T_k is contained in the image of $F(S \times (\sigma_k, 2))$ with $\sigma_k \to 2$, we see that T_k tends to the end facing S as $k \to \infty$. Since furthermore every T_k separates the end facing S from C' and $T_k \cup T_{k+1}$ bounds a submanifold homeomorphic to $S \times I$, every point in the component E of the complement of $\tilde{C}'$ facing T, which is a neighbourhood of the end facing S, lies between T_k and T_{k+1} for some k, where we regard T as T_0. Hence, this embedding is a surjection to the component E. This also shows that $\tilde{C}'$ is indeed a compact core and completes the proof of the proposition. □

Now we can prove Theorem 4.1.

Proof of Theorem 4.1. By applying Proposition 4.14 to every geometrically infinite end of $\mathbf{H}^2/G'$ by considering the subgroup Γ' corresponding to the image of the fundamental group of the boundary component of a compact core facing the end as before, we can see that every end of $\mathbf{H}^3/G'$ has a neighbourhood which is homeomorphic to the product of a closed surface and an open interval. (For a geometrically finite end has a neighbourhood with such a product structure, as is shown in Morgan [**33**] using the nearest point retraction to a small neighbourhood of the convex core.) Using the main result of Canary [**7**], Theorem 4.1 follows. □

REMARK 5. In the last step of the proof above, in fact we need not use the main theorem of Canary [**7**] itself. For each geometrically infinite end e, we have a sequence of pleated surfaces $\{f_{t_k}\}$ realizing weighted simple closed curves $\{\gamma_{t_k}\}$ converging to a measured lamination $\mu \in \mathcal{M}(S)$, which tends to the end e as $k \to \infty$. To prove Theorem 4.1, we only need to prove that there exists a constant $K > 0$ such that for every compression curve γ on S, the geodesic length of γ with respect to the hyperbolic structure induced by f_{t_k} is bounded by K below for every k as was explained in Thurston [**45**] and Canary [**7**]. This is shown in Affirmation 2.3.1 in Otal [**38**].

CHAPTER 5

Branched covers and geometric limit

Our proof of Theorem 6.1 uses Canary's construction of a negatively curved metric on a branched cover of a hyperbolic 3-manifold branched over a null-homologous simple closed curve in [**7**]. We shall recall results of Canary. The first observation that Canary made to prove the existence of negatively curved branched covering is the following lemma. (Lemma 5.2 in [**7**], and Lemma 3.1.2 in [**8**].)

LEMMA 5.1. *Let N be a compact irreducible 3-manifold. Let γ be a null-homologous collection of simple closed curves in* Int$N$ *intersecting all compressing discs for ∂N. Then any 2-fold or 3-fold branched cover of N branched over γ has incompressible boundary. For any component S of ∂N, its inverse image in the branched cover is a disjoint union of surfaces homeomorphic to S. Furthermore, in the case of 3-fold covering, the branched cover is also acylindrical.*

Having observed this, Canary used the following argument to prove his theorem (§5 of [**7**]). Let M be an almost compact hyperbolic 3-manifold which is regarded topologically as the interior of a compact 3-manifold $\overline{M}$. Take a null-homologous collection γ of simple closed curves in M intersecting all compressing discs for $\partial\overline{M}$. Let γ^* be the union of closed geodesics freely homotopic to γ. Suppose, for the moment, that γ^* has no singular points, i.e., each component is simple and the components are mutually disjoint. Let $\tilde{M}$ be a 3-fold branched cover of $\overline{M}$ branched over γ^*. (We shall consider only 3-fold cover from now on till the end of the chapter.) Using the result of Gromov-Thurston [**21**], Canary proved that Int$\tilde{M}$ admits a metric with sectional curvature ≤ -1 pinched below, which is obtained by perturbing near the inverse image $\tilde{\gamma}$ of γ^*, the singular hyperbolic structure obtained by pulling back the hyperbolic structure on M using the covering projection from Int$\tilde{M}$ to M. (Lemma 5.6 of [**7**].)

The pinching constant depends only on the number $\mathrm{Rad}^{\perp}(\gamma^*) = \inf_p\{\mathrm{length}(p)\}$, where p ranges over the homotopy classes of paths in M with endpoints on γ^* which are not homotopic into γ^*, and the "length" means the minimal length of the arcs representing the homotopy class. By Lemma 5.1, the 3-manifold $\tilde{M}$ has incompressible boundary. Canary proved that since Int$\tilde{M}$ has a pinched negatively curved metric, the theory of Bonahon in [**5**] can be applied by a straightforward generalization and every end of Int$\tilde{M}$ is simply degenerate. (§6 of [**7**])

Consider next the case when γ^* has singular points. Canary proved that, even in this case, a branched cover $\tilde{M}$ and the pinched negatively curved metric on its interior as desired can be constructed in the following manner. (Lemma 5.7 of [**7**].) For each singular point of γ^*, choose small disjoint balls centred at points on γ^*, one ball on each side of the singular point, e.g., for a double point, we take four balls. We perturb the original hyperbolic metrics in the balls to Riemannian metrics with pinched negative curvature in such a way that in the new metric, the closed geodesics homotopic to γ^* has no singular points. We can further assume that there exists a tubular neighbourhood of the new closed geodesic, denoted again by γ^*, in which the sectional curvature is constantly -1 after the metric perturbation. Then we can use the same technique as in the case when γ^* has no singular points, to obtain a branched cover $\tilde{M}$ with incompressible boundary such that Int$\tilde{M}$ admits a metric with pinched negative curvature.

Note that in this case, $\mathrm{Rad}^{\perp}\gamma^*$ for the closed geodesic γ^* with respect to the perturbed metric depends on the metric perturbation; hence the pinching constant also depends on the perturbation. However, $\mathrm{Rad}^{\perp}\gamma^*$ is bounded below if we perturb the metric in such a way that every two points on the new γ^* that are near in M are also near on the length metric (i.e., the metric with respect to the parameter) on the new γ^*. In other words, it is all right if we can perturb the metric in such a way that γ^* have no "near misses". We can control the perturbation in order not to cause new near misses on γ^*. In the case when the original γ^* has two points too near in M, which are not near on γ^*, we can use the same kind of local perturbation of the metric as we used for singular points, to remove such pairs of points. Thus we have shown the following.

LEMMA 5.2. *There exists a universal constant $\epsilon > 0$ depending only on a positive constant ϵ' with the following property. Let M be any almost hyperbolic 3-manifold and γ a null-homologous essential closed curve such that the length of the homotopic closed geodesic γ^* is greater than ϵ'. Then there is a metric deformation near γ^* as above such that $\mathrm{Rad}^{\perp}\gamma^* > \epsilon$ for the new γ^* with respect to the perturbed metric. Moreover under the same assumption, balls for a perturbation can be taken large enough and a perturbation in the balls can be chosen so as not to change the curvature too much.*

Hence for such M and γ^*, the pinching constant of the metric on $\tilde{M}$ can be bounded below by a universal negative constant.

Now, let us turn to the situation in which we are going to use the result above. Let Γ be a geometrically finite Kleinian group without parabolic elements such that $M = \mathbf{H}^3/\Gamma$ has a core homeomorphic to a compression body, and let $\{(\Gamma_i, \phi_i)\}$ be a sequence of quasi-conformal deformations of Γ with isomorphisms ϕ_i from Γ to Γ_i. We have a sequence of geometrically finite hyperbolic 3-manifolds $\{\mathbf{H}^3/\Gamma_i\}$ with homotopy equivalences $\Phi_i : M = \mathbf{H}^3/\Gamma \to \mathbf{H}^3/\Gamma_i$ homotopic to homeomorphisms, which induce isomorphisms conjugate to ϕ_i from Γ to Γ_i. Now let $\overline{M}$ be the natural compactification of

M, that is, the Kleinian manifold $(\mathbf{H}^3 \cup S^2_\infty)/\Gamma$, and similarly let $\overline{\mathbf{H}^3/\Gamma_i}$ be the Kleinian manifold $(\mathbf{H}^3 \cup S^2_\infty)/\Gamma_i$. We fix a null-homologous collection γ of simple closed curves in M, which intersects every compressing disc for $\overline{M}$. We shall construct a branched covering $\tilde{M}_i$ of $\overline{\mathbf{H}^3/\Gamma_i}$ branched over the union of the closed geodesics γ_i^* homotopic to $\Phi_i(\gamma)$, or disjoint simple closed curves near it when γ_i^* has singular points or near misses, for each i by the method above. As was explained above, $\mathrm{Int}\overline{\tilde{M}}_i$, which we shall denote by $\tilde{M}_i$ admits a metric with pinched negative sectional curvature. We specify the metric in the following way.

Consider the sequence $\{(\mathbf{H}^3/\Gamma_i, \gamma_i^*)\}$ and choose a base frame v_i on a basepoint x_i on γ_i^*. Let $(\mathbf{H}^3/\Gamma_\infty, \gamma_\infty^*, v_\infty)$ be the geometric limit in the sense of Gromov of a subsequence of $\{(\mathbf{H}^3/\Gamma_i, \gamma_i^*, v_i)\}$ (which we shall denote again by the same symbol) and let $\rho_i : B_{r_i}(\mathbf{H}^3/\Gamma_i, x_i) \to B_{r_i}(\mathbf{H}^3/\Gamma_\infty, x_\infty)$ be an approximate isometry associated with the geometric convergence where $r_i \to \infty$ as $i \to \infty$. (We need to take a subsequence of $\{\Gamma_i\}$ at several stages in our argument. We shall omit to mention it from now on because taking a subsequence does not change our situation at all.) Such a limit exists because the injectivity radii at points of γ_i^* are bounded below by a positive constant (by Margulis' lemma) independent of i as the diameters of the γ_i^*'s are bounded both above and below by positive constants as follows. As the sequence $\{\Gamma_i\}$ converges algebraically, the lengths and the distances from x_i of the components of γ_i^* are bounded above by a constant independent of i. Moreover, since the algebraic limit Γ' has no parabolic elements, the lengths of the components of γ_i^* are bounded below by a positive constant independent of i. Therefore we also see that the limit γ_∞^* is a collection of non-degenerate closed geodesics. However, γ_∞^* may have finitely many singular points. For each singular point of γ_∞^*, choose small disjoint balls $\{b_k\}$ centred at points on γ_∞^*, one ball on each side of the singular point as in the argument above.

LEMMA 5.3. *We can take a subsequence of* $\{\Gamma_i\}$ *so that for any* $\epsilon > 0$, *there exist* i_0 *and* $\delta > 0$ *satisfying the following.*

($\star$) For every $i > i_0$ and any pair x_1^i, x_2^i on γ_i^* that are connected by an arc essential relative to γ_i^* with length less than δ, there exists a singular point x^∞ on γ_∞^* such that $d_{\mathbf{H}^3/\Gamma_i}(\rho_i^{-1}(x^\infty), x_j^i) < \epsilon$ for $j = 1, 2$.

PROOF. Let σ denote the set of singular points of γ_∞^*. We consider the homotopy classes of arcs with endpoints on $\gamma_\infty^* - \sigma$ relative to $\gamma_\infty^* - \sigma$. For every positive number d there are only finitely many such homotopy classes that can be realized by arcs with length less than d. (For σ consists of finitely many points and the injectivity radii are bounded below by a positive constant on γ_∞^*.) Therefore, there exists a positive constant d' such that every arc having length less than d' either represents an inessential homotopy class or is homotopic relative to γ_∞^* to an arc passing through a singular point. On the other hand, for any small e, there is a constant e' such that every arc with length less than e' that represents an essential homotopy class as an arc with endpoints on $\gamma_\infty^* - \sigma$ but is relatively homotopic into γ_∞^* is contained in the e-neighbourhood of a singular point.

For sufficiently large i, any essential arc with endpoints on γ_i^* corresponds to an arc essential relative to $\gamma_\infty^* - \sigma$, hence to either an arc essential relative to γ_∞^* or an arc homotopic relative to γ_∞^* to an arc passing through a singular point. Therefore, for any ϵ, if we take a sufficiently small δ, every arc essential relative to γ_i^* with length less than δ is mapped by ρ_i into the ϵ-neighbourhood of a singular point. □

To resolve the singular points, we deform the metric of $\mathbf{H}^3/\Gamma_\infty$ in $\{b_k\}$ keeping the sectional curvature pinched above and below by negative constants, and get a negatively curved 3-manifold M_∞ and the disjoint union of simple closed geodesics γ_∞^+ with respect to the deformed metric, which is homotopic to γ_∞^* as was done by Canary. Fix a small constant $\epsilon > 0$, and we can take an integer i_0 (and $\delta > 0$) satisfying ($\star$). Then, the Riemannian metric on $\rho_i^{-1}(b_k)$ induced from b_k by $d\rho_i^{-1}$ has sectional curvature pinched above and below by negative constants independently of i for $i > i_0$ with some i_0, since the deformed metric on $\rho_i^{-1}(b_k)$ converges to that of b_k as $i \to \infty$ and $\rho_i|\rho_i^{-1}(b_k)$ approaches to an isometry. (Recall that $d\rho_i$ approaches an orthogonal map as $i \to \infty$).

DEFINITION 5.4. Let m_i be the metric on $\mathbf{H}^3/\Gamma_i$ after the perturbation, and let M_i be the Riemannian 3-manifold whose base topological manifold is $\mathbf{H}^3/\Gamma_i$, and which has the metric m_i. Let γ_i^+ be the closed geodesics homotopic to $\Phi_i(\gamma)$ with respect to m_i.

Evidently, M_i with this new metric m_i converges to M_∞ geometrically in the sense of Gromov as $i \to \infty$. Since γ_i^+ converges to γ_∞^+, the closed geodesics γ_i^+ are simple and disjoint for sufficiently large i. Construct a 3-fold branched cover (as a manifold, forgetting the metric) $\dot{M}_i$ of $\overline{\mathbf{H}^3/\Gamma_i}$ branched over γ_i^+ for such i. Let $\bar{M}_i$ denote the interior of $\dot{M}_i$ with the metric induced from (M_i, m_i) by the covering projection. Then, $\bar{M}_i$ has a structure of a negatively curved cone-manifold with singular locus $\tilde{\gamma}_i^+$. By the property ($\star$), the number $\mathrm{Rad}^\perp\gamma_i^+$ for the metric m_i is bounded below by a positive constant independent of i.

DEFINITION 5.5. We take a tubular neighbourhood V_∞ of γ_∞^+ which contains all the b_k. Let V_i by the pull-back of V_∞ by ρ_i, which is a tubular neighbourhood of γ_i^+ such that the diameter of V_i is bounded above and the diameter of its meridian is bounded below by positive constants independent of i. Let $\tilde{V}_i$ be the preimage of V_i in $\bar{M}_i$.

Since $M_i - V_i$ is hyperbolic, the covering translations act on $\bar{M}_i - \tilde{V}_i$ by hyperbolic isometries without fixed points.

LEMMA 5.6. *The action of* $\mathbf{Z}_3$ *on* $\bar{M}_i$ *associated with the branched covering* $p_i : \bar{M}_i \to M_i$ *converges geometrically (as metric spaces) in the sense of*

Gromov to an isometric action of $\mathbf{Z}_3$ *on a negatively curved cone-manifold* $\bar{M}_\infty$ *as* $i \to \infty$, *which gives rise to a branched covering* $p_\infty : \bar{M}_\infty \to M_\infty$ *branched over* γ_∞^+ *after taking a subsequence; that is, the action can be pushed forward to* $\bar{M}_\infty$ *by approximate isometries, and converges to an isometric* $\mathbf{Z}_3$*-action on* $\bar{M}_\infty$. *The tubular neighbourhoods* $\tilde{V}_i$ *converge to a tubular neighbourhood* $\tilde{V}_\infty$ *of the preimage of* γ_i^+ *in* $\bar{M}_\infty$, *and the quotient* $(\bar{M}_\infty - \tilde{V}_\infty)/\mathbf{Z}_3 = p_\infty(\bar{M}_\infty - \tilde{V}_\infty)$ *is isometric to the complement of an open tubular neighbourhood of* γ_∞^+ *in* $\mathbf{H}^3/\Gamma_\infty$.

PROOF. Since $\bar{M}_i - \tilde{V}_i$ is hyperbolic and covers $M_i - V_i$, it converges to a 3-fold cover $\hat{M}_\infty$ of $M_\infty - V_\infty$ by Theorem 8.3 in Paulin [**39**]. Each solid torus V_i has a branched cover $\tilde{V}_i$ with branching locus γ_i^+. It is obvious that $\tilde{V}_i$ with the induced metric as a cone-manifold converges geometrically as metric spaces (in the sense of Gromov) to a solid torus having a closed geodesic as cone singularity that is a 3-fold branched covering of V_∞ with branching locus γ_i^+. By pasting $\hat{M}_\infty$ and $\tilde{V}_\infty$ at their frontiers, we get a 3-fold branched cover $\bar{M}_\infty$ of M_∞ with branching locus γ_∞^+ which has a structure of negatively curved cone-manifold.

Since we assumed that $b_k \subset V_\infty$, we can see that $M_\infty - V_\infty$ is isometric to the complement of an open tubular neighbourhood of γ_∞^+ in $\mathbf{H}^3/\Gamma_\infty$. □

By the technique of Gromov-Thurston as used in Canary [**7**], we can endow a pinched negatively curved metric on $\tilde{M}_\infty$ deforming the metric on $\tilde{V}_\infty$, which coincides with the original metric in a thin neighbourhood of $\partial\tilde{V}_\infty$. We can also put a negatively curved metric on $\tilde{V}_i$ coinciding with the original metric in a thin neighbourhood of $\partial\tilde{V}_i$ which converges to geometrically to that of $\tilde{V}_\infty$. By replacing the metric in $\tilde{V}_i$ with this new one, we get a negatively curved manifold, which we denote by $\tilde{M}_i$. We summarize this as a lemma.

LEMMA 5.7. *There is a negatively curved* 3*-manifold* $\tilde{M}_i$ *topologically identified with the* 3*-fold branched cover* $\bar{M}_i$ *of* M_i, *such that the metric in* $\tilde{M}_i - \tilde{V}_i$ *coincides with the original metric on* $\bar{M}_i - \tilde{V}_i$. *The manifolds* $\tilde{M}_i$ *converge geometrically to a negatively curved* 3*-manifold* $\tilde{M}_\infty$ *which is homeomorphic to* $\bar{M}_\infty$, *and there is a solid torus* $\tilde{V}_\infty$ *in it such that* $\tilde{M}_\infty - \tilde{V}_\infty$ *is isometric to a* 3*-fold cover of* $\mathbf{H}^3/\Gamma_\infty - V_\infty$.

CHAPTER 6

Non-realizable measured laminations

Throughout this chapter, we consider a Kleinian group Γ without parabolic elements such that $\mathbf{H}^3/\Gamma$ has a compact core homeomorphic to a compression body, and its quasi-conformal deformations $\{(\Gamma_i, \phi_i)\}$. We assume in this chapter the the sequence $\{(\Gamma_i, \phi_i)\}$ converges strongly to (Γ', ϕ) which does not have parabolic elements. We shall prove that in this situation, the hyperbolic 3-manifold $\mathbf{H}^3/\Gamma'$ either is geometrically finite or satisfies the condition of Theorem 4.1, hence is almost compact in particular. After proving this, in the following two chapters, we shall prove that for a Kleinian group G and its quasi-conformal deformations as give in Theorem 2.1, a subgroup Γ of G corresponding to the image of the fundamental group of the compressible boundary component of a compact core satisfies the assumption above unless the limit set of Γ' is the entire S^2_∞.

The following is our main result in this chapter.

THEOREM 6.1. *Let Γ be a geometrically finite Kleinian group without parabolic elements such that $\mathbf{H}^3/\Gamma$ has a compact core which is a compression body. Let $\{(\Gamma_i, \phi_i)\}$ be a sequence of quasi-conformal deformations of Γ with isomorphisms $\phi_i : \Gamma \to \Gamma_i$. Suppose that $\{(\Gamma_i, \phi_i)\}$ converges strongly to a geometrically infinite Kleinian group (Γ', ϕ) and that Γ' has no parabolic elements. Let C' be a compact core of $\mathbf{H}^3/\Gamma'$. Let S be a component of $\partial C'$ facing a geometrically infinite end. Then there exists a sequence of simple closed curves $\{\delta_j\}$ on S as follows.*

(1) *The projective classes $\{[\delta_j] \in \mathcal{PL}(S)\}$ converge to a projective lamination $[\lambda]$ in the projectivized Masur domain $\mathcal{PM}(S)$. (We regard $\mathcal{PM}(S)$ as the entire $\mathcal{PL}(S)$ when S is incompressible in C'.)*

(2) *The closed geodesic δ_j^* which is homotopic to δ_j in $\mathbf{H}^3/\Gamma'$ tends to the end facing S as $j \to \infty$.*

It will be proved in the next chapter (Theorem 7.1) that a compact core C' of $\mathbf{H}^3/\Gamma'$ is homeomorphic to a compression body and that $\Phi|C$ is homotopic to a homeomorphism to C' in the situation of Theorem 6.1. Therefore *we prove the theorem assuming that C' is a compression body and there exists a homeomorphism $h' : C \to C'$ which is homotopic to $\Phi|C$.*

If S is a component of the interior boundary of C', Theorem 6.1 follows from Bonahon's theorem applied to a cover of $\mathbf{H}^3/\Gamma'$ associated to $\pi_1(S)$, and Theorem 9.2.2 of Thurston **[45]** (Thurston's covering theorem) which states that such a covering by a hyperbolic 3-manifold with a simply degenerate end is proper and finite sheeted (see also Canary [9] and Ohshika [35].) Thus we can assume that S is the exterior boundary of C'. Throughout this chapter, when we refer to the Masur domain of S, *we regard S as the boundary of the compression body C'.*

The gist of the proof of Theorem 6.1 is as follows. Let S_i be the boundary component of the convex core of $\mathbf{H}^3/\Gamma_i$ that is homotopic to $\Phi_i(S)$. Take a branched cover $\tilde{M}_i$ of $\mathbf{H}^3/\Gamma_i$ with a negatively curved metric as was explained in the last chapter. We shall consider the lift $\tilde{S}_i$ of S_i to $\tilde{M}_i$, and define a marked hyperbolic structure m_i on S induced from $\tilde{S}_i$. The limit of the hyperbolic structure $\{m_i\}$ in the Thurston compactification of the Teichmüller space $\mathcal{T}(S)$, after passing through a subsequence, will be shown to be a maximal projective lamination. This limit lamination, which plays the role of $[\lambda]$ in the statement of Theorem 6.1, will be proved to be contained in $\mathcal{PM}(S)$.

Now let us start the proof. Take a collection of null-homologous closed curves γ in $\mathbf{H}^3/\Gamma$, and construct branched covers $\tilde{M}_i$ of $\mathbf{H}^3/\Gamma_i$ as in chap. 5. In the present case, we can assume that γ is connected by the following reason. First observe that the set of projective classes of simple closed curves which are null-homologous on S is dense in $\mathcal{PL}(S)$. This follows from the fact that we can approximate every simple closed curve by a sequence of weighted null-homologous simple closed curves as we can see by using Dehn twists. Hence, in particular, using the facts that $\mathcal{M}(S)$ is open and that the weighted simple closed curves are dense in $\mathcal{ML}(S)$, we see that $\mathcal{M}(S)$ contains weighted null-homologous simple closed curves densely. Since C is a compression body, a simple closed curve in $\mathcal{M}(S)$ intersects every compression disc for ∂C. Thus we can choose γ, over which we are going to construct a branched covering, to be such a null-homologous simple closed curve in $\mathcal{M}(S)$.

Let S_i be a boundary component of the convex core C_{Γ_i} of $\mathbf{H}^3/\Gamma_i$ which is homotopic to $\Phi_i(S)$. Then S_i can be regarded as the image of a pleated surface $h_i : S \to S_i$ homotopic to $\Phi_i \circ \Phi^{-1}|S$. Here the homeomorphism h_i can be chosen uniquely up to homotopy in $\mathbf{H}^3/\Gamma'$, but not uniquely up to homotopy on S_i. We fix an h_i as above for the moment, and shall replace it afterward when necessary. The surface S_i has a hyperbolic metric induced from $\mathbf{H}^3/\Gamma_i$ as a path metric.

Fix a basepoint in $\mathbf{H}^3$ and a base-frame on it which are projected to basepoints $x_i \in \mathbf{H}^3/\Gamma_i$ and base-frames v_i on x_i so that the sequence $\{(\mathbf{H}^3/\Gamma_i, v_i)\}$ converges geometrically to $(\mathbf{H}^3/\Gamma', v_\infty)$ with a base-frame v_∞ in $\mathbf{H}^3/\Gamma'$. We denote the closed geodesic homotopic to $\Phi_i(\gamma)$ in $\mathbf{H}^3/\Gamma_i$ by γ_i^*, and its geometric limit by γ_∞^* as in the last chapter.

It is necessary that the boundary component S_i of the convex core does not intersect a tubular neighbourhood of the branching locus to be able to consider the hyperbolic structure on the boundary of the lift of the convex core. We shall prove first that we can choose an appropriate γ such that the closed geodesic γ_i^* has a tubular neighbourhood which does not intersect S_i.

LEMMA 6.2. *If $d(\gamma_i^*, S_i) \to 0$ as $i \to \infty$, then the pleated surfaces S_i converge geometrically to a pleated surface S_∞ in $\mathbf{H}^3/\Gamma'$ which contains γ_∞^* if we choose an appropriate basepoint on S_i.*

PROOF. Since $\{\Gamma_i\}$ converges algebraically, there is an upper bound for $d(x_i, \gamma_i^*)$, which is independent of i. We can assume that x_i lies on γ_i^* by conjugating the Γ_i's by conformal automorphisms bounded independently of i without changing the limit up to conjugacy in $PSL_2\mathbf{C}$. Then by assumption, $d(x_i, S_i)$ is bounded as $i \to \infty$. Let y_i be a point on S_i such that $d(x_i, y_i)$ is bounded as $i \to \infty$. To begin with, we shall prove by contradiction that the injectivity radius of S_i at y_i with respect to the hyperbolic metric on S_i induced from $\mathbf{H}^3/\Gamma_i$ does not go to 0 as $i \to \infty$.

Suppose, on the contrary, that the injectivity radius of S_i at y_i goes to 0, and let δ_i be an essential simple closed curve on S_i passing y_i whose length goes to 0 as $i \to \infty$. Since $\{\Gamma_i\}$ converges algebraically and $d(x_i, y_i)$ is bounded as $i \to \infty$, it is impossible that there is an essential loop in $\mathbf{H}^3/\Gamma_i$ passing y_i whose length goes to 0 as $i \to \infty$, by Jørgensen's inequality ([**23**]). Therefore δ_i is null-homotopic in $\mathbf{H}^3/\Gamma_i$ for sufficiently large i.

Because the length of δ_i goes to 0, we can take an embedded annulus A_i on S_i containing δ_i as its core such that its width goes to ∞ whereas the length of either component of ∂A_i goes to 0 as $i \to \infty$. Let α_i and β_i be the components of ∂A_i. Since the lengths of α_i and β_i go to 0, they bound embedded discs D_i^α and D_i^β respectively in the convex core C_{Γ_i}, whose diameters go to 0 as $i \to \infty$. We can assume that $\mathrm{Int}D_i^\alpha \cap S_i = \emptyset$ and $\mathrm{Int}D_i^\beta \cap S_i = \emptyset$ by pushing the discs into $\mathrm{Int}C_{\Gamma_i}$ by an isotopy fixing the boundary, and that $D_i^\alpha \cap D_i^\beta = \emptyset$ by interchanging sub-discs in finite steps to remove intersection. Let B_i be the ball in C_{Γ_i} that is bounded by the embedded 2-sphere $A_i \cup D_i^\alpha \cup D_i^\beta$. Then the minimum distance in B_i between the discs $d_{B_i}(D_i^\alpha, D_i^\beta)$ goes to infinity as $i \to \infty$ as we shall show in the following.

Since the core of A_i is null-homotopic in $\mathbf{H}^3/\Gamma_i$, there are no closed leaves in the pleating locus of S_i that are contained in A_i; for compact leaves in the pleating locus are closed geodesics in $\mathbf{H}^3/\Gamma_i$. Also the core of A_i cannot be homotoped into a complementary region of the pleating locus since any simple closed curve in a complementary region which is essential on S_i is also essential in $\mathbf{H}^3/\Gamma_i$. Therefore there exists a leaf of the pleating locus which enters A_i by α_i and exits A_i by β_i. (Otherwise A_i would contain a compact leaf of the pleating locus around which such a leaf spirals.) Let l_i be a geodesic segment on the leaf of the pleating locus as above in A_i, one of whose endpoints is on α_i and the other on β_i. By the construction of A_i, the length of l_i goes to infinity as $i \to \infty$. We can see that $d_{B_i}(D_i^\alpha, D_i^\beta)$ goes to infinity as $i \to \infty$: for, if $d_{B_i}(D_i^\alpha, D_i^\beta)$ did not go to infinity, we could connect the endpoints of l_i by joining three arcs in B_i whose lengths are bounded above as $i \to \infty$; the first one is on D_i^α, the second one is an arc with bounded length connecting D_i^β and D_i^β in B_i, and the last one is on D_i^β. Since the arc obtained by joining the three arcs above would be homotopic to l_i in B_i, this would contradict the fact that l_i is geodesic and its length goes to infinity. Similarly we can see that $d_{B_i}(y_i, D_i^\alpha)$ and $d_{B_i}(y_i, D_i^\beta)$ go to infinity.

Now, the points x_i and y_i can be joined in C_{Γ_i} by an arc with length bounded as $i \to \infty$ since $d(x_i, y_i)$ is bounded, and $x_i \in C_{\Gamma_i}$. Because $d(y_i, D_i^\alpha)$ and $d(y_i, D_i^\beta)$ go to ∞ as $i \to \infty$, the geodesic arc joining x_i and y_i as above and the basepoint x_i are contained in B_i. Also, the closed geodesic γ_i^* must intersect either D_i^α or D_i^β when it goes out of B_i starting from x_i. Note that it cannot exit B_i by only one of D_i^α and D_i^β because if it could, the endpoints of this part of γ_i^* in B_i containing x_i lie both on either D_i^α or D_i^β and are very near on it as the diameters of D_i^α and D_i^β go to 0. This cannot happen because γ_i^* is geodesic, B_i is simply connected, and the length of the part of γ_i^* above goes to infinity since $d(D_i^\alpha, x_i)$ and $d(D_i^\beta, x_i)$ go to infinity. Therefore each component of $\gamma_i^* \cap B_i$ has one endpoint on D_i^α and the other on D_i^β.

The length of the component of $\gamma_i^* \cap B_i$ containing x_i goes to infinity since $d_{B_i}(D_i^\alpha, D_i^\beta)$ goes to infinity. Hence the length of γ_i^* also goes to infinity, which contradicts the assumption that Γ_i converges algebraically and γ_i^* is homotopic to $\Phi_i(\gamma)$. Thus we have proved that the injectivity radius of S_i at y_i does not go to 0 as $i \to \infty$.

As the injectivity radius of S_i at y_i does not go to 0, the pleated surfaces S_i with basepoint at y_i converge to a pleated surface S_∞ in the geometric limit $\mathbf{H}^3/\Gamma_\infty$, which is equal to $\mathbf{H}^3/\Gamma'$ by assumption (see Thurston [**48**] and Canary-Epstein-Green [**10**].) On the other hand, since we assumed that $d(\gamma_i^*, S_i) \to 0$, we have $d(\gamma_\infty^*, S_\infty) = 0$. Since S_i is a boundary component of the convex core, it is bent along its pleated locus only to one direction with respect to the normal vectors and closed geodesics exist only one side of S_i. Since S_∞ is their geometric limit, it has the same properties. Hence γ_∞^* can intersect S_∞ only when S_∞ contains γ_∞^*.

Note moreover that the limit surface S_∞ above is also a boundary component of the convex core of $\mathbf{H}^3/\Gamma'$. This can be shown by the same argument as the proof of Theorem 2.1 in Ohshika [**35**]: it can be proved that the surface S_∞ must be separating and closed geodesics exist only one side of S_∞ by using approximate isometries.

We shall show that even when the situation of Lemma 6.2 occurs (i.e., $d(\gamma_i^*, S_i) \to 0$,) we can change γ to a new one, which is still connected, so as to avoid the situation of Lemma 6.2.

LEMMA 6.3. *We can choose a connected γ with required properties such that $d(\gamma_i^*, S_i)$ does not go to 0 as $i \to \infty$.*

PROOF. Let λ^1, λ^2 be connected maximal measured laminations in the Masur domain of S, whose supports are distinct. Then every component of $S - \lambda^1 \cup \lambda^2$ is an open disc. Since the set of projective classes of null-homologous simple closed curves is dense in $\mathcal{PL}(S)$, we can choose a null-homologous simple closed curve χ^j in the Masur domain of S whose projective class is close to that of λ^j for $j = 1, 2$ such that every component of $S - (\chi^1 \cup \chi^2)$ is an open disc.

Let $\chi^j(i)$ be the closed geodesic in $\mathbf{H}^3/\Gamma_i$ which is freely homotopic to $\Phi_i(\chi_j)$. Let $\chi^j(\infty) \subset \mathbf{H}^3/\Gamma_\infty = \mathbf{H}^3/\Gamma'$ be the geometric limit of $\chi^j(i)$ as $i \to \infty$. Suppose that the situation of Lemma 6.2 occurs, whichever of $\chi^1(i)$ and $\chi^2(i)$ we let γ_i^* be. Then the limit pleated surface S_∞ contains both $\chi^1(\infty)$ and $\chi^2(\infty)$. Since S_∞ is a geometric limit of the pleated surfaces S_i, unless it is homeomorphic to S, it has a puncture corresponding to a cusp with respect to the two-dimensional hyperbolic metric on S_∞. Since every component of $S - (\chi^1 \cup \chi^2)$ is an open disc, the length of either $\chi^1(i)$ or $\chi^2(i)$ goes to infinity if S_∞ has a cusp. Therefore S_∞ is homeomorphic to S, and is totally geodesic since it contains $\chi^1(\infty)$ and $\chi^2(\infty)$ and $S-(\chi^1 \cup \chi^2)$ is an open disc.

As was remarked just before this lemma, the surface S_∞ is a boundary component of the convex core of $\mathbf{H}^3/\Gamma_\infty$. A boundary component of the convex core is isotopic to a boundary component of compact core by the relative core theorem by McCullough [**28**], and the uniqueness of compact core up to homeomorphism proved by McCullough-Miller-Swarup which was cited in §1.B. Since the only component of C' that is homeomorphic to S is S itself, this implies that S_∞ is isotopic to S. It follows that S_∞ is compressible since S is the exterior boundary component of C'. This is a contradiction as compressible surfaces cannot be totally geodesic. Hence for at least one of χ^1 and χ^2 the situation of Lemma 6.2 does not happen. □

Choose γ as in Lemma 6.3, and from now on, we assume that $d(S_i, \gamma_i^*)$ does not go to 0 as $i \to \infty$. Then we can also assume that for the closed geodesic γ_i^+ in the perturbed metric m_i defined in Definition 5.4, $d(\gamma_i^+, S_i)$ does not go to 0 either, and that $p_i(V_i) \cap S_i = \emptyset$ for sufficiently large i by choosing the balls b_k for deforming the metric and the deformation of the metric there to be smaller if necessary. Thus we can lift S_i to an incompressible embedded surface $\tilde{S}_i$ in $\tilde{M}_i$, which is isometric to S_i. The closed geodesics γ_i^+ converge geometrically to a closed geodesic γ_∞^+ in the geometric limit M_∞ which is isometric to $\mathbf{H}^3/\Gamma'$ outside a tubular neighbourhood of γ_∞^+.

Recall that we fixed a homeomorphism h_i temporarily. We are now in position to change it to a suitable one for proving the existence of simple closed curves projectively converging inside the Masur domain, which will represent an "ending lamination". (We shall not define the term "ending lamination" precisely, which is irrelevant in this paper.) In the following lemma, we shall prove that homeomorphisms h_i can be chosen to be homotopic to pleated surfaces g_i that converge geometrically to a pleated surface homotopic to the inclusion of S in $\mathbf{H}^3/\Gamma'$, outside a tubular neighbourhood of the branching locus γ_i^+. These will turn out to be essential properties for proving Theorem 6.1.

PROPOSITION 6.4. *We can choose a homeomorphism $h_i : S \to S_i$ homotopic in $\mathbf{H}^3/\Gamma_i$ to $\Phi_i \circ \Phi^{-1}|S$ such that there exists a pleated surface $g_i : S \to \mathbf{H}^3/\Gamma_i$ homotopic to $\Phi_i \circ \Phi^{-1}|S$ realizing a measured lamination in $\mathcal{M}(S)$ as follows.*

(1) *The image of g_i is contained in $\mathbf{H}^3/\Gamma_i - N(\gamma_i^+)$, where $N(\gamma_i^+)$ denotes the open tubular neighbourhood of γ_i^+, which is equal to the tubular neighbourhood V_i in M_i defined in Definition 5.5 under the identification of M_i with $\mathbf{H}^3/\Gamma_i$ forgetting the metrics. (Recall that V_i is covered by the solid torus $\tilde{V}_i$ in $\tilde{M}_i$ outside of which the metric is intact.)*
(2) *The pleated surface g_i is homotopic to h_i in $\mathbf{H}^3/\Gamma_i - N(\gamma_i^+)$.*
(3) *The minimum distance from the basepoint x_i to $g_i(S)$ is bounded as $i \to \infty$.*
(4) *The sequence of pleated surfaces $\{g_i\}$ converges geometrically, after taking a subsequence, to a pleated surface $g_\infty : S \to \mathbf{H}^3/\Gamma'$ (as unmarked pleated surfaces) homotopic to the inclusion of S if for each i, we choose a base-frame on a point on S which is mapped to a point within distance bounded independently of i from x_i.*

Before starting the proof of Proposition 6.4, we should observe the following property of pleated surfaces homotopic to S_i outside $N(\gamma_i^+)$.

LEMMA 6.5. *Fix a homeomorphism $h_i : S \to S_i$ homotopic to $\Phi_i \circ \Phi^{-1}|S$ in $\mathbf{H}^3/\Gamma_i$. There is a positive constant ζ independent of i and h_i with the following property. If a pleated surface $f : S \to \mathbf{H}^3/\Gamma_i$ is homotopic to h_i in $\mathbf{H}^3/\Gamma_i - N(\gamma_i^+)$, then any essential simple closed curve γ on S such that $f(\gamma)$ is null-homotopic has length at least ζ with respect to the hyperbolic metric on S induced by f from $\mathbf{H}^3/\Gamma_i$.*

PROOF. Suppose, seeking a contradiction, that such a constant ζ does not exist. Then, after taking a subsequence, there are an essential simple closed curve c_i on S and a pleated surface $f_i : S \to \mathbf{H}^3/\Gamma_i$ homotopic to h_i in $\mathbf{H}^3/\Gamma_i - N(\gamma_i^+)$ with null-homotopic $f_i(c_i)$ such that the length of c_i with respect to the hyperbolic metric on S induced by f_i goes to 0 as $i \to \infty$. This implies that there is a singular disc D_i bounded by $f_i(c_i)$ in $\mathbf{H}^3/\Gamma_i$ whose diameter goes to 0 as $i \to \infty$.

Recall that every compressing disc for S_i intersects γ_i^+ essentially. Since $f_i(c_i)$ is homotopic to the boundary of a compressing disc for S_i outside $N(\gamma_i^+)$, the disc D_i also intersects γ_i^+ essentially. Since γ_i^+ is null-homologous, D_i cannot intersect γ_i^+ at one point. Therefore, there is an arc on D_i connecting two points of $D_i \cap \gamma_i^+$ that is not homotopic into γ_i^+. Since the diameter of D_i goes to 0, we have that $\mathrm{Rad}^{\perp}(\gamma_i^+)$ also goes to 0. This contradicts our construction of γ_i^+. □

PROOF OF PROPOSITION 6.4. [**Construction of** g_i.] Fix h_i as before for the moment. Let λ_i be a measured lamination on S obtained by pulling back the bending lamination of S_i by h_i.

If the minimum distance from $h_i(S)$ to x_i is uniformly bounded, we let g_i be h_i, and proceed to the part [property of g_i]. We assume from now on until the end of this part that the minimum distance from $h_i(S)$ to x_i, hence also to γ_i^+ goes to infinity as $i \to \infty$.

We can extend the sequence $\{(\Gamma_i, \phi_i)\}$ to a continuous one-parameter family of quasi-conformal deformations $\{(\Gamma_t, \phi_t)\}$ $(t \in [0, \infty))$ of Γ. We shall define a pleated surface $h_t : S \to \mathbf{H}^3/\Gamma_t$ homotopic in $\mathbf{H}^3/\Gamma_t$ to $\Phi_t \circ \Phi^{-1}|S$ whose image is a boundary component of the convex core, and a measured lamination λ_t corresponding to the bending locus of h_t, by the same way as for h_i, in the following manner. In the first place, note that there exists a K-bi-Lipschitz diffeomorphism $L_{t,t'} : \mathbf{H}^3/\Gamma_t \to \mathbf{H}^3/\Gamma_{t'}$ homotopic to $\Phi_{t'} \circ \Phi_t^{-1}$ where $K \to 1$ as $t \to t'$. The diffeomorphism $L_{t,t'}$ approaches to an isometry as $t \to t'$. In particular, there is a constant $\epsilon(t, t')$ going to 0 as $t \to t'$ such that for any closed geodesic c in $\mathbf{H}^3/\Gamma_t$, the geodesic curvature of $L_{t,t'}$ goes to 0. (See Douady-Earles [**12**] and Thurston [**45**].)

Let S_t be the unique boundary component of the convex core of $\mathbf{H}^3/\Gamma_t$ that is compressible, which is homotopic to $\Phi_t \circ \Phi^{-1}(S)$. (Note that for integer parameters, this definition is compatible with that of S_i before.) We can see as follows that , the image of the convex core of $C(\mathbf{H}^3/\Gamma_t)$ by the map $L_{t,t'}$ converges to the convex core of $C(\mathbf{H}^3/\Gamma_t)$ with respect to the Chabauty topology.

Suppose, seeking a contradiction, that there are $\delta > 0$ and points x_t in S_t which are mapped by $L_{t,t'}$ to the outside of the δ-neighbourhood of the convex core of $\mathbf{H}^3/\Gamma_{t'}$. First, assume that x_t lies on a maximal measured lamination realized on S_t. Approximating the maximal measured lamination by weighted simple closed curves, we see that there is a closed geodesic c_t with length bounded uniformly from below by a positive constant at the minimum distance less than $\delta/4$ from x_t. Since the geodesic curvature of $L_{t,t'}(c_t)$ goes to 0 as $t \to t'$, for t sufficiently close to t', the closed geodesic c_t^* homotopic to $L_{t,t'}(c_t)$ in $\mathbf{H}^3/\Gamma_{t'}$ stays in the ϵ_t-neighbourhood of $L_{t,t'}(c_t)$ with $\epsilon_t \to 0$ as $t \to t'$. Since $L_{t,t'}(c_t)$ is within the minimum distance $\delta/2$ from $L_{t,t'}(x_t)$ for every t sufficiently near to t', we can find t close to t' such that the closed geodesic c_t^* does not lie inside the convex core of $\mathbf{H}^3/\Gamma_{t'}$. This is a contradiction for the convex core contains every closed geodesic.

For general x_t, there is a complementary region R_t of the maximal measured lamination, which contains x_t. By our assumption of reductio ad absurdum, $L_{t,t'}(x_t)$ lies outside the δ-neighbourhood of the convex core. On the other hand, $L_{t,t'}(R_t)$ converges to either a totally geodesic surface with geodesic boundary or a geodesic line with respect to the Chabauty topology as $t \to t'$. The frontier of the limit of $L_{t,t'}(R_t)$ as $t \to t'$ is contained in the convex core. Since the δ-neighbourhood of the convex core is convex, the limit of $L_{t,t}(R_t)$ is also contained in the convex core. Hence, any point in $L_{t,t'}(R_t)$ lies in the δ-neighbourhood for t close to t'. This is a contradiction. Thus we have proved that $L_{t,t'}$ maps the convex core of $\mathbf{H}^3/\Gamma_t$ into that of $\mathbf{H}^3/\Gamma_{t'}$ for t close to t'.

By the same argument interchanging the roles of t and t', we see that the image of the convex core of $\mathbf{H}^3/\Gamma_t$ by $L_{t,t'}$ converges to the convex core of $\mathbf{H}^3/\Gamma_{t'}$. In particular $L_{t,t'}(S_t)$ is contained in a arbitrarily small neighbourhood of $S_{t'}$ for t sufficiently near to t'.

We choose $\{h_t\}$ so that $L_{t,t'} \circ h_t$ is homotopic to $h_{t'}$ in a thin tubular neighbourhood of $S_{t'}$ for every t, t'. Redefine h_i to be equal to h_t above for each t corresponding to an integer i.

LEMMA 6.6. *The measured lamination λ_t varies continuously with respect to t.*

PROOF. As argued above, $L_{t,t'}(S_t)$ converges to S_t' with respect to the Chabauty topology as $t \to t'$. The bending lamination is determined by the minimal total curvatures of isotopy classes of simple closed curves on the surface S when they are mapped on S_t. Let c be a simple closed curve on S. Then the total curvature of $L_{t,t'}(\Phi_t(c))$ converges to that of $\Phi_{t'}(c)$ by the convergence of $L_{t,t'}(S_t)$ to $S_{t'}$. Therefore the minimal total curvature of an isotopy class on S_t also converges to that on $S_{t'}$. Hence the bending lamination λ_t converges to $\lambda_{t'}$. □

The measured lamination λ_t might not be contained in the Masur domain $\mathcal{M}(S)$. Still, we can prove the following.

LEMMA 6.7. *For every t, t', the measured lamination λ_t can be realized by a pleated surface in $\mathbf{H}^3/\Gamma_{t'}$ homotopic to $h_{t'}$.*

PROOF. By definition, $(\Gamma_{t'}, \phi_{t'})$ is a quasi-conformal deformation of Γ_t. The measured lamination λ_t is realized by h_t on S_t in $\mathbf{H}^3/\Gamma_t$. We are going to prove that the realizability of measured laminations are invariant by quasi-conformal deformations. Let λ_t' be a maximal geodesic lamination containing λ_t which is realized by the pleated surface h_t.

The pleated surface $h_t : S \to \mathbf{H}^3/\Gamma_t$ is lifted to a map $\tilde{h}_t : \mathbf{H}^2 \to \mathbf{H}^3$, which is a pleated surface bent along the lift $\tilde{\lambda}_t \subset \mathbf{H}^2$ of the geodesic lamination λ_t', and is equivariant with respect to the action of $\pi_1(S)$ on $\mathbf{H}^2$ and that of Γ_t on $\mathbf{H}^3$, i.e., we have $\tilde{h}_t(\gamma x) = (\phi_t \circ \iota(\gamma))\tilde{h}_t(x)$, where $\iota : \pi_1(S) \to \Gamma$ is the homomorphism induced from the inclusion of S into $\mathbf{H}^3/\Gamma$.

Recall that we have a bi-Lipschitz homeomorphism $L_{t,t'} : \mathbf{H}^3/\Gamma_t \to \mathbf{H}^3/\Gamma_{t'}$. This map $L_{t,t'}$ is lifted to a bi-Lipschitz homeomorphism $\tilde{L}_{t,t'} : \mathbf{H}^3 \to \mathbf{H}^3$ which is equivariant with respect to the action of Γ_t and $\Gamma_{t'}$ under the isomorphism $\phi_{t'} \circ \phi_t^{-1}$. On the geodesic lamination $\tilde{\lambda}_t$, we define a continuous map $k|\tilde{\lambda}_t : \tilde{\lambda}_t \to \mathbf{H}^3$ which is equivariant with respect the action of $\pi_1(S)$ on $\tilde{\lambda}_t$ as covering translations and that of $\Gamma_{t'}$ on $\mathbf{H}^3$ such that for each leaf ℓ of $\tilde{\lambda}_t$, its image is the geodesic whose endpoints at infinity are equal to those of $\tilde{L}_{t,t'}(\tilde{h}_t(\ell))$, where we regard $\tilde{L}_{t,t}$ as acting on S^2_∞ by a quasi-conformal homeomorphism. The existence of such a map is easy to prove by using the equivariance of $\tilde{h}_t$ and $\tilde{L}_{t,t'}$. Since every complementary region of $\tilde{\lambda}_t$ is an ideal triangle by the maximality of λ'_t, we can extend $k|\tilde{\lambda}_t$ to the entire hyperbolic plane equivariantly, which gives rise to an equivariant pleated surface $k : \mathbf{H}^2 \to \mathbf{H}^3$ realizing $\tilde{\lambda}_t$. By taking a quotient, we get a pleated surface realizing λ_t homotopic to $h_{t'}$. □

By the lemma above, in the hyperbolic 3-manifold $\mathbf{H}^3/\Gamma_t$, we have a family of pleated surfaces $\{h_t^s\}(s \in [1,t])$ homotopic to h_t, such that h_t^s realizes λ_s.

This family has the following property.

Claim 4. After taking a subsequence for $\{\Gamma_i\}$, one of the following holds for the family of pleated surfaces $\{h_i^s\}$.

(1) $\inf_{s\in[1,i]} d(x_i, h_i^s(S))$ goes to infinity as $i \to \infty$.
(2) There are a constant K independent of i, and $u_i \in [1,i]$ for each i such that $h_i^{u_i}$ is homotopic to h_i in $\mathbf{H}^3/\Gamma_i - N(\gamma_i^+)$ and the image of the pleated surface $h_i^{u_i}(S)$ is within the minimum distance K from x_i.

Proof. Suppose that the first condition does not hold. Then, after taking a subsequence, there exist a constant K' independent of i, and a parameter $s_i \in [1,i]$ for each i such that $h_i^{s_i}(S)$ is within the minimum distance K' from x_i. Since γ_i^+ converges to a closed curve in $\mathbf{H}^3/\Gamma'$, there is a constant C independent of i bounding the diameters of the γ_i^+ from above. We define a constant K_0 to be $K' + C$. By Lemma 6.5, there is a positive constant ζ independent of i bounding from below the lengths of simple closed geodesics on S with respect to the hyperbolic metric induced by $h_i^{s_i}$ whose image by h_i are null-homotopic in $\mathbf{H}^3/\Gamma_i - N(\gamma_i^+)$. Fix a positive number ϵ smaller than the (three-dimensional) Margulis constant and this ζ. Note that the diameters modulo the (two-dimensional) ϵ-thin part of hyperbolic surfaces homeomorphic to S are bounded and the ϵ-thin part of the S are mapped into the ϵ-thin part of $\mathbf{H}^3/\Gamma_i - N(\gamma_i^+)$ by $h_i^{s_i}$ as we took ϵ less than ζ. Therefore, we can see that the diameters modulo the three-dimensional ϵ-thin part of the images of pleated surfaces homotopic to h_i in $\mathbf{H}^3/\Gamma_i - N(\gamma_i^+)$ are bounded by a constant independent of i. Considering the geometric limit of $\mathbf{H}^3/\Gamma_i$ and the pleated surfaces, and using Margulis' lemma and the fact that the geometric limit has no parabolic elements, we can see that the diameters themselves of such pleated surfaces within the minimum distance K_0 from x_i are also bounded by a constant independent of i, which we denote by D. Similarly, there is a constant D' such that any such pleated surface within the minimum distance $K_0 + D$ from x_i has diameter less than or equal to D'. Let $t_i \in [1, i]$ be the infimum of $t \in [1, i]$ such that for every $s \in [t, i]$, the pleated surface h_i^s is homotopic to h_i in $\mathbf{H}^3/\Gamma_i - N(\gamma_i^+)$. If $t_i \leq s_i$, then the second condition of our claim holds by letting u_i be s_i. Thus we can assume that $t_i > s_i$, in particular $t_i > 1$.

If there is an $r_i \in (t_i, i]$ such that $h_i^{r_i}(S)$ is within the minimum distance $K_0 + D + D'$ from x_i for infinitely many i, then the second condition of our claim holds by letting K be $K_0 + D + D'$. Hence we can also assume that for all i and $s \in (t_i, i]$, the surface $h_i^s(S)$ is apart from x_i at the minimum distance greater than $K_0 + 2D$. Now consider the pleated surface $h_i^{t_i}$. First we shall show that the image $h_i^{t_i}(S)$ is at a minimum distance greater than $K_0 + D$ from x_i.

Take a monotone decreasing sequence of parameters $\{\sigma_j\}$ converging to t_i. By Lemma 6.5, using the same argument as Otal's proof of Lemma 4.2, we can see that as $j \to \infty$, the sequence $\{h_i^{\sigma_j}\}$ converges uniformly to a pleated surface $h_i^* : S \to \mathbf{H}^3/\Gamma_i$ realizing the geometric limit of the supports of λ_{σ_j} if we take a suitable subsequence. By Lemma 4.9 and the continuity of λ_t with respect to t proved in Lemma 6.6, the limit geodesic lamination contains the support of λ_{t_i}. Hence h_i^* realizes the measured lamination λ_{t_i}. Since the image of λ_{t_i} does not depend on pleated surfaces realizing it, as one can see by the same argument as the proof of Claim 1, the images of h_i^* and $h_i^{t_i}$ have non-empty intersection. By our definition of D', we see that $h_i^{t_i}(S)$ is apart from x_i at a minimum distance greater than $K_0 + D$.

By the same argument as Thurston's construction of homotopy consisting of negatively curved surfaces which we explained in chap. 4, a homotopy between h_i^* and $h_i^{t_i}$ consisting of pleated surfaces and negatively curved surfaces can be constructed, each of which contains the realization of λ_{t_i} in its image. (Here we should observe that for each component U of $S \setminus \lambda_{t_i}$, the restriction $h_i^{t_i}|U$ is incompressible since $h_{t_i}|U$ is incompressible by the definition of bending lamination. This makes it possible to construct a homotopy without the assumption that λ_{t_i} is contained in the Masur domain.) Since the diameters of the images of negatively curved surfaces homotopic to h_i in $\mathbf{H}^3/\Gamma_i - N(\gamma_i^+)$ within the minimum distance $K_0 + D$ from x_i are bounded by the same constant D' as the pleated surfaces, we can see in particular none of the surfaces in the homotopy can touch γ_i^+. This means that h_i^* and $h_i^{t_i}$ are homotopic on $\mathbf{H}^3/\Gamma_i - N(\gamma_i^+)$.

Next consider a monotone increasing sequence $\{s_j'\}$ converging to t_i. Then by the same argument as above, we can see that for sufficiently large j, the pleated surface $h_i^{s_j'}$ is homotopic to $h_i^{t_i}$ in $\mathbf{H}^3/\Gamma_i - N(\gamma_i^+)$. This implies that there is a positive constant ϵ such that for any $s \in [t_i - \epsilon, i]$, the pleated surface h_i^s is homotopic to h_i in $\mathbf{H}^3/\Gamma_i - N(\gamma_i^+)$. This contradicts our definition of t_i. Thus we have shown if the first condition of our claim fails to hold, then the second condition holds. □

If the second condition holds, we let g_i be $h_i^{u_i}$ and proceed to the part [property of g_i]. Suppose now that the first condition of the claim holds: that $\inf_{s\in[1,i]} d(x_i, h_i^s(S))$ goes to infinity as $i \to \infty$. Then by the same argument as above, we can see that for sufficiently large i, the pleated surface h_i^s is homotopic to h_i in $\mathbf{H}^3/\Gamma_i - N(\gamma_i^+)$ for every $s \in [1, i]$.

Consider the measured lamination λ_1, which is the bending lamination of h_1. First suppose that λ_1 has a complementary region which is not simply connected. Then there is a simple closed curve γ such that $i(\gamma, \lambda_1) = 0$. This implies that $\gamma \cup \lambda_1$ is realized by h_1 on S_1. By Lemma 6.7, for each i, there is a pleated surface g_i^1 realizing $\gamma \cup \lambda_1$. The images of pleated surfaces $h_i^1(S)$ and $g_i^1(S)$ have non-empty intersection. It follows, by the same argument as before, that the minimum distance between x_i and $g_i^1(S)$ also goes to infinity as $i \to \infty$. Let γ_i be the closed geodesic in $\mathbf{H}^3/\Gamma_i$ homotopic to $h_i(\gamma)$, which we know to be contained in $g_i^1(S)$. As $\{\Gamma_i\}$ converges strongly to Γ', the geodesics γ_i converge geometrically to the closed geodesic homotopic to γ in $\mathbf{H}^3/\Gamma'$. It follows that the minimum distance from x_i to γ_i, hence to $g_i^1(S)$ is bounded as $i \to \infty$. This is a contradiction. Thus we have proved that every complementary region of λ_1 is simply connected. We can further see the following.

CLAIM 5. The measured lamination λ_1 is contained in the Masur domain $\mathcal{M}(S)$.

PROOF. Suppose that λ_1 is not contained in the Masur domain. Then, by definition, there is a measured lamination $\nu \in \overline{\mathcal{C}}(S)$ such that $i(\nu, \lambda_1) = 0$. Since every complementary region of λ_1 is simply connected, this implies that the supports of ν and λ_1 coincide.

As $\nu \in \overline{\mathcal{C}}(S)$, there exists a sequence of weighted simple closed curves $\{w_j e_j\}$ converging to ν such that e_j is the boundary of a compressing disc for C'. By Lemma 4.9, after taking a subsequence, we can assume that the simple closed curves e_j converge in the geodesic lamination space to a geodesic lamination ν' which contains the support of ν. Since the support of ν is that of λ_1 and every complementary region of λ_1 is mapped by h_1 to a totally geodesically, this geodesic lamination ν' is realized by h_1 on S_1.

By Lemma 4.8, there exists an $\eta > 0$ depending only on ν', and for any $\epsilon > 0$, there exist a train track τ carrying ν' and a continuous map $h : S \to \mathbf{H}^3/\Gamma_i$ homotopic to h_1, which is adapted to a tied neighbourhood N_τ of τ satisfying the following two conditions: the image of each branch of τ has length greater than η, and for each two adjacent branches s_1, s_2 of τ, the exterior angle formed by $h(s_1)$ and $h(s_2)$ is less than ϵ. Since $h(e_j)$ is null-homotopic, there exists a homotopy between $h|\gamma_i$ to a constant map to one point $*$, which we denote by $H : (D^2, 0) \to (\mathbf{H}^3/\Gamma_1, *)$, where 0 denotes the centre of D^2.

Since $\{e_j\}$ converges geometrically to ν', which is carried by τ, the simple closed curve e_j is carried by τ for sufficiently large j. We can triangulate D^2 by triangles one of whose vertex is on 0 and the others are on ∂D^2.

We can choose a homotopy H such that each triangle with respect to this triangulation is mapped homeomorphically to a geodesic triangle. Then the map H induces a hyperbolic metric on D^2 with a singularity at the point 0 such that ∂D^2 is a piece-wise geodesic. In this situation, we can use the same argument as the proof of Proposition 5.1 in Bonahon [**5**] as follows, which we already used in the proof of Lemma 4.10. At each point z in ∂D^2 corresponding to the image of $e_j \cap N_\tau$, consider a perpendicular with respect to the induced metric on D^2. For a given $\delta > 0$, if there exists a perpendicular issued at z with length δ, then we let it be λ_z. Otherwise, i.e., if the perpendicular reaches a point in ∂D^2 within distance δ, then we let the perpendicular with the maximal length be λ_z.

By the Gauss-Bonnet theorem, if we take a sufficiently small δ independently of j, there exists a $\delta' > 0$ depending on δ and the bound ϵ of the exterior angles (such that $\delta' \to 0$ as $\epsilon \to 0$ fixing δ) as follows. If a λ_z intersects another $\lambda_{z'}$ with $z \neq z'$, then the distance between z and z' on ∂D^2 is less than δ', and there is exactly one vertex corresponding to a switch of τ between z and z' (on the side with shorter length.)

Let $B \subset \partial D^2$ be the set of points z for which the above occurs with some z'. Then we have $\text{length}(B) \leq 2\delta'\text{length}(\partial D^2)/\eta$. The union of perpendiculars λ_z ($z \in \partial D^2 - B$) has area greater than $\delta\{\text{length}(\partial D^2) - \text{length}(B)\}$. On the other hand the area of D^2, which is equal to the area of $H(D^2)$, is bounded above by $\epsilon\text{length}(\partial D^2)/\eta - 2\pi < \epsilon\text{length}(\partial D^2)/\eta$ by the Gauss-Bonnet theorem. Thus we have $\text{length}(B) \geq (1 - \epsilon/\delta\eta)\text{length}(\partial D^2)$. Combining this inequality with the precedent one, we get $2\delta'/\eta \geq (1 - \epsilon/\delta\eta)$. As was seen before, fixing δ and making ϵ go to 0, the constant δ' goes to 0. This contradicts the inequality above. □

Since $\mathcal{M}(S)$ is open, the set of weighted simple closed curves is dense in $\mathcal{M}(S)$. Also since $\mathcal{M}(S)$ is invariant under the positive scalar multiplication, there is a simple closed curve γ in $\mathcal{M}(S)$ which can be joined with λ_1 by an arc in $\mathcal{M}(S)$. Since Γ_i is geometrically finite, there is a pleated surface $k_i : S \to \mathbf{H}^3/\Gamma_i$ homotopic to h_i which realizes γ. By the technique which we employed in chap. 4, we can construct a homotopy $H_i : S \times I \to \mathbf{H}^3/\Gamma_i$ between h_i^1 and k_i consisting of pleated surfaces and negatively curved surfaces. Since $k_i(S)$ is within a uniformly bounded minimum distance from x_i whereas the minimum distance between x_i and $h_i^1(S)$ goes to infinity as $i \to \infty$, we can see that for some t_i the surface $H_i(\ ,t_i)$ is within uniformly bounded minimum distance from x_i and homotopic to h_i in $\mathbf{H}^3/\Gamma_i - N(\gamma_i^+)$. (The latter condition is satisfied if we choose t_i so that every surface $H_i(\ ,t)$ for $t \in [t_i, 1]$ is disjoint from $N(\gamma_i^+)$.) Since any point on the interpolated surfaces is within a uniformly bounded minimum distance from the original pleated surfaces, we can assume this surface $H(\ ,t_i)$ to be a pleated surface. We let g_i be such a surface $H(\ ,t_i)$ then.

[**Property of** g_i.] Thus in any case, we have a pleated surface $g_i : S \to \mathbf{H}^3/\Gamma_i$ homotopic to h_i in $\mathbf{H}^3/\Gamma_i - N(\gamma_i^+)$ which lies within a minimum distance bounded independently of i from x_i.

It remains to show that the sequence of pleated surfaces $\{g_i\}$ converges geometrically after taking a subsequence to a pleated surface $g_\infty : S \to \mathbf{H}^3/\Gamma'$ homotopic to the inclusion. First we shall show that the pleated surfaces $g_i : S \to \mathbf{H}^3/\Gamma_i$ converge geometrically to a pleated surface $g'_\infty : S \to \mathbf{H}^3/\Gamma'$. (Recall that $\mathbf{H}^3/\Gamma'$ is equal to the geometric limit $\mathbf{H}^3/\Gamma_\infty$ by assumption.) Let y_i be a point on $g_i(S)$ such that $d(x_i, y_i)$ is bounded as $i \to \infty$, which is guaranteed to exist by the definition of g_i. Conjugating Γ_i and $\Gamma' = \Gamma_\infty$ by bounded conformal automorphisms, regard y_i as a basepoint on which lies a base frame associated with the geometric convergence $\mathbf{H}^3/\Gamma_i \to \mathbf{H}^3/\Gamma_\infty$. The sequence of pleated surfaces $\{g_i\}$ converges geometrically to a pleated surface $\hat{g}_\infty : S' \to \mathbf{H}^3/\Gamma_\infty = \mathbf{H}^3/\Gamma'$, where S' is a subsurface of S such that its frontier $\mathrm{Fr}S'$ consists of essential simple closed curves on S (of course it is possible that $S = S'$.) This follows from the compactness of unmarked pleated surfaces. The surface S' is a geometric limit of $\{(S, g_i)\}$ with a base frame on the basepoint corresponding to y_i. Hence S' admits a hyperbolic, and S' differs from S if and only if it has a cusp. Note that we can apply the compactness to our situation because the injectivity radius at y_i with respect to the hyperbolic structure on S induced by g_i is bounded below by a positive constant since g_i can be lifted to an incompressible pleated surface in $\tilde{M}_i$ isometrically and in $\tilde{M}_i$, the injectivity radius at a lift of y_i is bounded below as was seen in chap. 5. (Or equivalently, because of Lemma 6.5.)

Now, consider the hyperbolic metric μ'_∞ on S' induced by $\hat{g}_\infty$ from $\mathbf{H}^3/\Gamma_\infty$. We claim that S' is the entire S in fact.

CLAIM 6. The surface S' defined above must coincide with the entire S.

PROOF. Suppose, seeking a contradiction, that $S \neq S'$. Let μ_i be the hyperbolic metric on S induced by g_i from $\mathbf{H}^3/\Gamma_i$. A simple closed curve c in S' homotopic to a component of $\mathrm{Fr}S'$ in S represents a parabolic cusp with respect to μ'_∞, which implies that there exists an arbitrarily short simple closed curve homotopic to c on S'. It follows that $\hat{g}_\infty(c)$ represents either the trivial element or a parabolic conjugacy class of Γ'. As we assumed that Γ' has no parabolic elements, $\hat{g}_\infty(c)$ must be null-homotopic. This implies that for any $\epsilon > 0$, there exists i_0 such that if $i > i_0$ then there exists an essential simple closed curve c_i on S such that the length of $g_i(c_i)$ is less than ϵ, which corresponds to c by an approximate isometry between (S, μ_i) and (S', μ'_∞). As the latter condition also implies that $g_i(c_i)$ is null-homotopic for large i, this contradicts Lemma 6.5. □

Hence S' must be equal to S, and the pleated surfaces g_i converge geometrically to a pleated surface $g'_\infty : S \to \mathbf{H}^3/\Gamma'$. Next we shall show that the limit pleated surface can be taken to be homotopic to the inclusion of S.

Let μ_∞ be the hyperbolic metric on S induced by g'_∞ from $\mathbf{H}^3/\Gamma'$. Since the pleated surfaces g_i converge to g'_∞ geometrically, for sufficiently large i, after passing through a subsequence, there exists an approximate isometry $\overline{\rho}_i : (S, \mu_i) \to (S, \mu_\infty)$ such that $\{\rho_i \circ g_i \circ \overline{\rho}_i^{-1}\}$ converges to g'_∞ uniformly, where $\rho_i : B_{r_i}(\mathbf{H}^3/\Gamma_i, x_i) \to B_{r_i}(\mathbf{H}^3/\Gamma', x_\infty)$ is an approximate isometry associated with the strong convergence $\Gamma_i \to \Gamma'$. In particular, $\rho_i \circ g_i \circ \overline{\rho}_i^{-1}$ is homotopic to g'_∞ for every sufficiently large i.

Take a basepoint x_S on S and let $\zeta_1, \ldots, \zeta_m$ be loops representing a generator system of $\pi_1(S, x_S)$. We can assume that $g'_\infty(x_S) = y_\infty$, where y_∞ is the limit of the basepoints $\{y_i\}$, which were assumed to be on $g_i(S)$, and that $g_i \circ \overline{\rho}_i^{-1}(x_S) = y_i$. Also by homotoping Φ_i, we can assume that there is a point $y_0 \in \mathbf{H}^3/\Gamma$ such that $\Phi_i(y_0) = y_i$. For each ζ_j, consider the sequence of homotopy classes $\{[\Phi_i^{-1} \circ \rho_i^{-1} \circ g'_\infty(\zeta_j)]\}$ with basepoint at y_0. Since Γ' is the algebraic limit of $\{\Gamma_i\}$, for sufficiently large i, we have $[\rho_i^{-1}(g'_\infty(\zeta_j))] = \phi_i \circ \phi^{-1} \circ g'_{\infty\#}([\zeta_j])$. Furthermore, there exists an element η_j of Γ such that $\phi(\eta_j) = g'_{\infty\#}[\zeta_j]$. Hence by taking a subsequence, we can assume that the homotopy class $[\Phi_i^{-1}\rho_i^{-1} \circ g'_\infty(\zeta_j)]$ is constant. By repeating this argument for each $j = 1, \ldots, m$, we can assume that $\Phi_i^{-1} \circ \rho_i^{-1} \circ g'_\infty(\zeta_1) (\simeq \Phi_i^{-1} \circ g_i \circ \overline{\rho}_i^{-1}(\zeta_1)), \ldots, \Phi_i^{-1} \circ \rho_i^{-1} \circ g'_\infty(\zeta_m) (\simeq \Phi_i^{-1} \circ g_i \circ \overline{\rho}_i^{-1}(\zeta_m))$ represent constant homotopy classes with basepoint at y_0 regardless of i in $\mathbf{H}^3/\Gamma$. Therefore there exists a continuous map $k : S \to \mathbf{H}^3/\Gamma$ such that $k \simeq \Phi_i^{-1} \circ g_i \circ \overline{\rho}_i^{-1}$ for every large i. On the other hand, as $g_i \simeq \Phi_i \circ \Phi^{-1}|S$, we have $\iota \circ \overline{\rho}_i^{-1} \simeq \Phi \circ k$, where $\iota : S \to \mathbf{H}^3/\Gamma'$ denotes the inclusion. Thus we have $\iota \simeq \Phi \circ k \circ \overline{\rho}_i \simeq \Phi \circ \Phi_i^{-1} \circ \rho_i^{-1} \circ g'_\infty \circ \overline{\rho}_i$. Since for any loop c in $\mathbf{H}^3/\Gamma'$, the loop $\rho_i^{-1}(c)$ is homotopic to $\Phi_i \circ \Phi^{-1}(c)$ for sufficiently large i, we have $\iota(\zeta_j) \simeq g'_\infty \circ \overline{\rho}_i(\zeta_j)$ for each ζ_j. Fix a sufficiently large i_0 and let $g_\infty : S \to \mathbf{H}^3/\Gamma'$ be $g'_\infty \circ \overline{\rho}_{i_0} : (S, \overline{\rho}_{i_0}^*(\mu_\infty)) \to \mathbf{H}^3/\Gamma'$. Then $\{g_i\}$ converges to g_∞ geometrically as unmarked pleated surfaces and g_∞ is homotopic to the inclusion ι. □

Next we shall define a sequence of marked hyperbolic structures $\{m_i\}$ on S. Let $\tilde{h}_i$ be a lift of $h_i : S \to S_i$ to $\tilde{M}_i$ whose image we denote by $\tilde{S}_i$. Since we fixed the surface $\tilde{S}_i$ before, the image of $\tilde{h}_i$ is uniquely determined. Our choice of $\tilde{h}_i$ as a map, however, has ambiguity even up to homotopy because the surface S_i, which is the projection of $\tilde{S}_i$, is compressible. To define $\tilde{h}_i$ without ambiguity up to homotopy, we specify a lift of h_i as follows. After choosing one sequence $\{g_i\}$ as in Lemma 6.4, let $\tilde{g}_i : S \to \tilde{M}_i$ be a lift of g_i which is homotopic to a homeomorphism onto $\tilde{S}_i$. Then $\{\tilde{g}_i\}$ converges geometrically, as unmarked pleated surfaces, to a lift $\tilde{g}_\infty$ of g_∞ to $\tilde{M}_\infty$. Since $\tilde{M}_\infty$ is a geometric limit of $\{\tilde{M}_i\}$ and the limit is compatible with the action of $\mathbf{Z}_3$ as was seen in chap. 5, there exists an approximate isometry $\tilde{\rho}_i : B_{r_i}(\tilde{M}_i, \tilde{x}_i) \to B_{r_i}(\tilde{M}_\infty, \tilde{x}_\infty)$ such that $\rho_i \circ p_i = p_\infty \circ \tilde{\rho}_i$, where ρ_i is an approximate isometry associated to the geometric convergence of $\{(\mathbf{H}^3/\Gamma_i, v_i)\}$ to $(\mathbf{H}^3/\Gamma', v_\infty)$.

DEFINITION 6.8. *Fixing a map* $\tilde{g}_\infty : S \to \tilde{M}_\infty$ *as above, we retake a map* $\tilde{g}_i : S \to \tilde{M}_i$ *with the same image as the original* $\tilde{g}_i$ *so that* $\tilde{\rho}_i \circ \tilde{g}_i \simeq \tilde{g}_\infty$ *in* $\tilde{M}_\infty - \tilde{V}_\infty$. Since $\tilde{\rho}_i \circ \tilde{g}_i \simeq \tilde{g}_\infty \circ \overline{\rho}_i$ in $\mathbf{H}^3/\Gamma_i - N(\gamma_i^+)$ for sufficiently large i, we can obtain the new $\tilde{g}_i$ only by changing the marking of the old one. *We also redefine* g_i *to be* $p_i \circ \tilde{g}_i$ *which has the same image as the old one.*

Then $g_i \simeq p_i \circ \tilde{\rho}_i^{-1} \circ \tilde{g}_\infty \simeq \rho_i^{-1} \circ g_\infty \simeq \rho_i^{-1} \circ \iota$; hence g_i is homotopic to $\Phi_i \circ \Phi^{-1}|S$ for sufficiently large i. (We can assume that this is true for all i by taking a subsequence.) Therefore this new g_i satisfies all the conditions of Proposition 6.4 by replacing the homeomorphism h_i from S to S_i if necessary. Note also that the geometric limit g_∞ of $\{g_i\}$ remains unchanged. (We shall denote this new h_i by $\overline{k}_i$ below.)

As $\tilde{g}_i$ is homotopic to a homeomorphism to $\tilde{S}_i$ (in $\tilde{M}_i - V_i$), the map $\tilde{g}_i : S \to \tilde{M}_i$ determines a homotopy equivalence $k_i : S \to \tilde{S}_i$ uniquely up to homotopy as maps to $\tilde{S}_i$ because $\tilde{S}_i$ is incompressible in $\tilde{M}_i$ and $\tilde{M}_i$ is not a surface bundle over a circle. (Refer for instance to §5.2 of Canary-Epstein-Green [**10**] for a proof of the uniqueness.) Let $\overline{k}_i : S \to \mathbf{H}^3/\Gamma_i$ be a pleated surface whose image is S_i such that $\overline{k}_i = p_i \circ k_i$. Then $\overline{k}_i$ is homotopic to g_i in $\mathbf{H}^3/\Gamma_i - N(\gamma_i^+)$, hence also to $\Phi_i \circ \Phi^{-1}|S$ in $\mathbf{H}^3/\Gamma_i$.

On the other hand, by lifting the hyperbolic structure on S_i, we have a hyperbolic structure on $\tilde{S}_i$. Hence we can define a marked hyperbolic structure m_i on S from the hyperbolic structure on $\tilde{S}_i$ and the homotopy equivalence k_i.

LEMMA 6.9. *The sequence of marked hyperbolic structures* $\{m_i\}$*, defined above, has no subsequences that converge inside the Teichmüller space* $\mathcal{T}(S)$.

PROOF. Suppose, seeking a contradiction, that a subsequence of $\{m_i\}$, which we denote again by $\{m_i\}$, converges inside $\mathcal{T}(S)$. Let ζ be a simple closed curve on S which represents a non-trivial homotopy class in $\mathbf{H}^3/\Gamma'$. Let ζ_i^* be the closed geodesic in $\mathbf{H}^3/\Gamma_i$ freely homotopic to $\Phi_i \circ \Phi^{-1}(\zeta)$, and let ζ_i^+ be the closed geodesic on S_i with respect to the two-dimensional hyperbolic structure on S_i, which is freely homotopic to $\overline{k}_i(\zeta)$. Since $\overline{k}_i$ is homotopic to $\Phi_i \circ \Phi^{-1}|S$ in $\mathbf{H}^3/\Gamma_i$, the two closed curves ζ_i^* and ζ_i^+ are freely homotopic in $\mathbf{H}^3/\Gamma_i$.

Because $\{\phi_i\}$ converges algebraically to ϕ and Γ' has no parabolic elements, the lengths of the closed geodesics ζ_i^* are bounded both above and below by positive constants independent of i. On the other hand, as we assumed that $\{m_i\}$ converges inside $\mathcal{T}(S)$, the lengths of the closed curves ζ_i^+ are also bounded both above and below by positive constants independent of i. This implies that the distance between ζ_i^* and ζ_i^+ is bounded above as $i \to \infty$. From this it follows that the sequence of pleated surfaces $\overline{k}_i : S \to \mathbf{H}^3/\Gamma_i$ stays in a bounded minimum distance from the basepoint x_i as $i \to \infty$ because $\overline{k}_i(S)$ contains ζ_i^+, which is within bounded distance from ζ_i^*, and the distance between ζ_i^* and the basepoint x_i is bounded above. Since furthermore the moduli of the S_i's are within a compact set of the moduli space by the assumption that $\{m_i\}$ converges inside $\mathcal{T}(S)$, the pleated surfaces $\overline{k}_i : S \to \mathbf{H}^3/\Gamma_i$ converge geometrically to a pleated surface $\overline{k}_\infty : S \to \mathbf{H}^3/\Gamma'$, as unmarked pleated surfaces.

Next we shall show that $\overline{k}_\infty(S)$ is a boundary component of the convex core of $\mathbf{H}^3/\Gamma'$. The argument which we use to prove this is the same as one in the proof of Theorem 2.1 in Ohshika [**35**]. (We also used a similar argument before Lemma 6.3.) Let us review it briefly here. The surface $\overline{k}_\infty(S)$ may not be embedded. Still it is approximated by an embedded surface arbitrarily closely because $\overline{k}_i(S)$ is embedded for every i. Suppose that $\overline{k}_\infty(S)$ is non-separating; that is, an embedding sufficiently closely approximating $\overline{k}_\infty(S)$ is non-separating. Then there exists a simple closed curve intersecting an approximating surface for $\overline{k}_\infty(S)$ at exactly one point transversely. It follows that there exists a simple closed curve whose algebraic intersection number with $\overline{k}_i(S)$ is 1 for sufficiently large i. This contradicts the fact that $\overline{k}_i(S)$ is separating (because $\overline{k}_i(S)$ is a boundary component of the convex core of $\mathbf{H}^3/\Gamma_i$.) A similar argument shows that there cannot exist a closed geodesic intersecting $\overline{k}_\infty(S)$ transversely and that there exists a side of $\overline{k}_\infty(S)$ such that the component of the complement of $\overline{k}_\infty(S)$ on that side intersects no closed geodesics. This implies that $\overline{k}_\infty(S)$ is a boundary component of the convex core of $\mathbf{H}^3/\Gamma'$, and is embedded.

Now we shall prove that there is a homotopy between g_i and $\overline{k}_i$ whose diameter is bounded as $i \to \infty$. Let x_S be a basepoint on S. Let $\{c_1, \ldots, c_m\}$ be loops representing a generator system for $\pi_1(S, x_S)$. Since $\{m_i\}$ is bounded in $\mathcal{T}(S)$, we can assume that the length of the closed geodesic loop c^i_j based at $\overline{k}_i(x_S)$ on S_i with respect to the metric m_i, which is homotopic to $\overline{k}_i(c_j)$ relative to $\overline{k}_i(x_S)$, is bounded as $i \to \infty$. On the other hand, since $\{\rho_i \circ g_i\}$ converges uniformly to g_∞, (note that in Definition 6.8 we have redefined g_i as such,) we can see that the length of the geodesic loop based at $g_i(x_S)$ homotopic to $g_i(c_j)$ on the surface $g_i(S)$ is bounded as $i \to \infty$ because ρ_i is an approximate isometry and the geodesic loops converge to one homotopic to $g_\infty(c_j)$ with basepoint at $g_\infty(x_S)$. We can assume that $g_i(c_j)$ is itself a geodesic loop on the pleated surface with respect to the two-dimensional hyperbolic metric by composing an isotopy on S. Since both $\overline{k}_i$ and g_i are homotopic to $\Phi_i \circ \Phi^{-1}|S$ as was seen before, we can homotope $\cup_j \overline{k}_i(c_j)$ to $\cup_j g_i(c_j)$ by a homotopy whose diameter is bounded as $i \to \infty$. It is easy to see that this homotopy extends to one between g_i and $\overline{k}_i$ preserving the condition that the diameter is bounded because $S - \cup_j c_j$ is topologically an open disc.

Thus we can see, by pushing forward the homotopy between g_i and $\overline{k}_i$ to $\mathbf{H}^3/\Gamma'$ by an approximate isometry for sufficiently large i, that the geometric limit $\overline{k}_\infty$ of $\{\overline{k}_i\}$ is homotopic to g_∞ which is the geometric limit of $\{g_i\}$, hence to the inclusion of S. This implies that the end of $\mathbf{H}^3/\Gamma'$ facing S is geometrically finite, and contradicts our assumption that the end is geometrically infinite. □

Since $\{m_i\}$ does not converge inside $\mathcal{T}(S)$ after taking a subsequence, there is a further subsequence that converges to a projective lamination $[\lambda]$ in the Thurston compactification of the Teichmüller space.

As was asserted in Proposition 4.7, Bonahon proved that for a measured lamination λ on S with an incompressible map $f : S \to M$ to a hyperbolic 3-manifold, one of the two alternative conditions is satisfied corresponding to whether or not λ can be realizable in M by a pleated surface. Canary proved in [**8**] that this proposition can be generalized to the case when M is a 3-manifold with pinched negatively curvature. Let us state his result more precisely. Recall that, as in hyperbolic 3-manifolds, we say that a continuous map h from hyperbolic surface S to a negatively curved 3-manifold N is adapted to a tied neighbourhood N_τ of a train track τ on S when h maps each branch of τ to a geodesic arc in N and each tie of N_τ to a point. Also note that there is a minor change in the statement from the original. In the alternative (2) of our statement, we consider the neighbourhood of $h'(c \cap N_\tau)$ instead of $h'(c)$ for our later convenience. Since $\text{length}(c - N_\tau)/\text{length}(c)$ goes to 0 as c approaches to λ, this change makes no difference to the proof.

LEMMA 6.10 (Canary). *Let N be a closed Riemannian 3-manifold with pinched negative sectional curvature. Let S be a closed hyperbolic surface. Let $h : S \to N$ be a continuous map which induces a monomorphism between the fundamental groups. Let α be a measured lamination on S. Then one of the following two conditions is satisfied, and they are mutually exclusive. Furthermore, the alternatives do not depend on the transverse measure of α, i.e., they are determined only by the support of α.*

(1) *For any $\epsilon > 0$, there exists a continuous map h' homotopic to h which is adapted to a tied neighbourhood of a train track carrying α such that* $\text{length}(h'(\alpha)) < \epsilon$.
(2) *For any $\epsilon > 0$ and $0 < t < 1$, there exist a train track τ carrying α and a continuous map h' homotopic to h which is adapted to a tied neighbourhood N_τ of τ and satisfies the following. Suppose that c is a simple closed curve on S such that $c/\text{length}_S(c)$ is sufficiently close to $\alpha/\text{length}_S(\alpha)$ in the measured lamination space $\mathcal{ML}(S)$. Let c^* be the closed geodesic which is homotopic to $h'(c)$ in N. Then there exists a part of c^* with length at least $t\text{length}_N(h'(c \cap N_\tau))$ which lies in the ϵ-neighbourhood of $h'(c \cap N_\tau)$.*

We shall prove that our measured lamination λ, to whose projective class $\{m_i\}$ converges, and the map $g_\infty : S \to \tilde{M}_\infty$ satisfy the condition (1) in Lemma 6.10.

PROPOSITION 6.11. *In Lemma 6.10, if we let N, h and α be respectively our limit branched covering $\tilde{M}_\infty$, $\tilde{g}_\infty$, and any measured lamination with the same support as λ, then the alternative (1) holds.*

We shall prove this proposition by contradiction, pulling back the situation in the limits $\mathbf{H}^3/\Gamma'$ and $\tilde{M}_\infty$ to $\mathbf{H}^3/\Gamma_i$ and $\tilde{M}_i$. For that, we need the following lemma, which states that if the condition (2) holds true for a measured lamination α on S and $\tilde{g}_\infty$, then it also does for α and $\tilde{g}_i$ for sufficiently large i.

LEMMA 6.12. *Suppose that in $\tilde{M}_\infty$, the condition (2) in Lemma 6.10 is satisfied replacing N, h and α with $\tilde{M}_\infty$, $\tilde{g}_\infty$, and λ. Then for any $\epsilon > 0$ and $0 < t < 1$, there exist an integer i_0, a train track τ carrying α, and a continuous map $h'_i : S \to \mathbf{H}^3/\Gamma_i$ adapted to a tied neighbourhood N_τ of τ for $i \geq i_0$, which is homotopic to $\tilde{g}_i$ and satisfies the following.*

(1) *As $i \to \infty$, $\text{length}(h'_i(\alpha))$ does not go to 0.*
(2) *If c is a simple closed curve on S such that $c/\text{length}_S(c)$ is sufficiently close to $\alpha/\text{length}_S(\alpha)$ in $\mathcal{ML}(S)$, then for each $i \geq i_0$, the closed curve $h'_i(c)$ is homotopic to a closed geodesic c'_i which has a part of length at least $t\text{length}_{\tilde{M}_i}(h'_i(c \cap N_\tau))$ lying in the ϵ-neighbourhood of $h'_i(c \cap N_\tau)$.*

PROOF OF LEMMA 6.12. Recall that there exists a $(1+\epsilon_i, r_i)$-approximate isometry $\tilde{\rho}_i : B_{r_i}(\tilde{M}_i, \tilde{x}_i) \to B_{r_i}(\tilde{M}_\infty, \tilde{x}_\infty)$, where $\epsilon_i \to 0$ and $r_i \to \infty$ as $i \to \infty$.

Let a be a geodesic arc in $B_{r_i}(\tilde{M}_\infty, \tilde{x}_\infty)$. Then $\tilde{\rho}_i^{-1}(a)$ is approximated arbitrarily finely by a piecewise geodesic arc whose total length is bounded by $(1+\epsilon_i)\text{length}(a)$, which implies that $\text{length}(\tilde{\rho}_i^{-1}(a)) \leq (1+\epsilon_i)\text{length}(a)$. Let a_i be the geodesic arc that is homotopic to $\tilde{\rho}_i^{-1}(a)$ fixing the endpoints. We can assume that a_i is contained in $B_{r_i}(\tilde{M}_i, \tilde{x}_i)$ for sufficiently large i. Then $\text{length}(a_i) \leq \text{length}(\tilde{\rho}_i^{-1}(a)) \leq (1+\epsilon_i)\text{length}(a)$. By applying the same argument to a_i instead of a, we have $\text{length}(a) \leq (1+\epsilon_i)\text{length}(a_i)$. Thus, we have $(1+\epsilon_i)^{-1}\text{length}(a) \leq \text{length}(a_i) \leq (1+\epsilon_i)\text{length}(a)$.

Next let a, b be two geodesic arcs in $B_{r_i}(\tilde{M}_\infty, \tilde{x}_\infty)$ which start from a common endpoint o. Let θ be the exterior angle formed by a and b at o. Let a_i and b_i be the geodesic arcs in $\tilde{M}_i$ homotopic to $\tilde{\rho}_i^{-1}(a)$ and $\tilde{\rho}_i^{-1}(b)$ respectively fixing the endpoints. Let θ_i be the exterior angle at $\tilde{\rho}_i^{-1}(o)$ formed by a_i and b_i. Then one can easily show by the hyperbolic trigonometry and the inequality in the last paragraph that there exists a sequence $\{\delta_i\}$ of positive real numbers going to 0 as $i \to \infty$ such that $|\theta - \theta_i| \leq \delta_i$.

Now, recall that in the proof of Proposition 5.1 in Bonahon [**5**] (Proposition 4.7 in the present paper) and its generalization by Canary (Lemma 6.10), it was proved that the alternative (2) holds true (unless (1) holds) by showing that for a given train track T carrying α, one can make a train track τ carried by T and a continuous map $\hat{h}_\infty$ adapted to a tied neighbourhood N_τ of τ in such a way that the total curvature and the quadratic variation of angle ("la variation angulaire totale" and "la variation quadratique angulaire" defined in [**5**]) of $\hat{h}_\infty(\alpha)$, where α is regarded as contained in N_τ, are arbitrarily small keeping the length of $\hat{h}_\infty(\alpha)$ away from 0. For such a map $\hat{h}_\infty : S \to \tilde{M}_\infty$, which is adapted to N_τ, we can make a map

$\hat{h}_i : S \to \tilde{M}_i$ such that for any branch b of τ, its image $\hat{h}_i(b)$ is the geodesic arc homotopic to $\tilde{\rho}_i^{-1}(\hat{h}_\infty(b))$ fixing the endpoints. Then by the results in the last paragraph, for any $\delta > 0$, there exists i_0 such that if $i > i_0$, then the total curvature and the quadratic variation of angle of $\hat{h}_i(\alpha)$ are bounded by those of $\hat{h}_\infty(\alpha)$ plus δ respectively, and such that the length of $h_i(\alpha)$ is bounded below by the length of $\hat{h}_\infty(\alpha)$ minus δ.

As we assumed that the alternative (2) held true for α and $\tilde{g}_\infty$, for any $\epsilon > 0$, we can take an $\hat{h}_\infty$ homotopic to $\tilde{g}_\infty$, which is adapted to a tied neighbourhood N_τ, such that the total curvature and the quadratic variation of angle of $\hat{h}_\infty(\alpha)$ are less than ϵ, while the length of $\hat{h}_\infty(\alpha)$ is kept away from 0. Since $\tilde{\rho}_i \circ \tilde{g}_i$ is homotopic to $\tilde{g}_\infty$, it follows from the remark in the last paragraph that for any $\epsilon > 0$, there exist i_0 and a map $\hat{h}_i$ homotopic to $\tilde{g}_i$ for $i > i_0$, which is adapted to the tied neighbourhood N_τ, such that the total curvature and the quadratic variation of angle of $\hat{h}_i(\alpha)$ are less than ϵ, and such that the length of $\hat{h}_i(\alpha)$ is bounded below by a positive constant independent of ϵ and i.

There was another constant which came into Bonahon's proof, the one denoted by c_1 in Lemme 5.10 in [**5**]. This c_1 is a constant depending on ϵ such that for any geodesic arc a on S with length less than ϵ, the geodesic arc homotopic to $\hat{h}_\infty(a)$ fixing the endpoints has length at most c_1. By using approximate isometries, it is easy to see that we can find a constant c_1 with this property for the maps $\hat{h}_i$ independently of i.

Since these are the only properties that were used to prove the condition (2) in the original proof of Proposition 4.7, i.e., Proposition 5.1 in [**5**], we have completed the proof of our lemma. □

PROOF OF PROPOSITION 6.11. Since the alternatives of Lemma 6.10 do not depend on the transverse measure of α, we have only to prove Proposition 6.11 under the assumption that $\alpha = \lambda$. Suppose, seeking a contradiction, that the alternative (2) in Lemma 6.10 is valid replacing N, h and α with $\tilde{M}_\infty$, $\tilde{g}_\infty$, and λ respectively. Then by Lemma 6.12, for any sufficiently large i, there exists a map h'_i homotopic to $\tilde{g}_i$, which is adapted to a tied neighbourhood N_τ carrying λ, such that the condition (2) in Lemma 6.10 holds true replacing N, h' and α with $\mathbf{H}^3/\Gamma_i$, h'_i, and λ respectively.

On the other hand, since $\{m_i\}$ converges to $[\lambda]$ in the Thurston compactification of the Teichmüller space, as was shown in Theorem 2.2 in Thurston [**49**], whose proof can be found in the argument of Thurston [**51**], there exists a sequence of weighted simple closed curves $\{\omega_i \xi_i\}$ converging to λ such that $\omega_i \mathrm{length}_{m_i}(\xi_i)$ goes to 0. Let ξ_i^* be the closed geodesic in $\tilde{M}_i$ freely homotopic to $\tilde{g}_i(\xi_i)$. Then $\omega_i \mathrm{length}(\xi_i^*)(\leq \omega_i \mathrm{length}_{m_i}(\xi_i))$ also goes to 0 as $i \to \infty$.

By the condition (2) in Lemma 6.10, for a fixed small $\epsilon > 0$ and t close to 1, a part of length at least $t \mathrm{length} h'_i(\xi_i \cap N_\tau)$ of the closed geodesic ξ_i^* must stay in the ϵ-neighbourhood of $h'_i(\xi_i \cap N_\tau)$ if i is sufficiently large. On the other hand, since length$(h_i'(\lambda))$ is bounded below by a positive constant, and the weight system on τ which $\omega_i\xi_i$ induces converges to the one which λ induces, we can see that $t\omega_i$length$(h_i'(\xi_i \cap N_\tau))$ is also bounded below by a positive constant as $i \to \infty$. Therefore $\omega_i$length(ξ_i^*) is also bounded below by a positive constant. This contradicts the fact that $\omega_i$length(ξ_i^*) goes to 0, which was shown above. □

LEMMA 6.13. *The measured lamination λ is connected and maximal: that is, for any simple closed curve c on S, the geometric intersection number $i(\lambda, c)$ is not* 0.

PROOF. We shall prove this lemma by contradiction. Suppose that there exists a simple closed curve c on S such that $i(c, \lambda) = 0$. We divide the argument into two cases: (i) the case when there exists a closed geodesic c^* in $\tilde{M}_\infty$ homotopic to $\tilde{g}_\infty(c)$; and (ii) the case when $\tilde{g}_\infty(c)$ represents a parabolic element, i.e., for any $\epsilon > 0$ there exists a closed curve freely homotopic to $\tilde{g}_\infty(c)$ with length less than ϵ. (Note that $\tilde{g}_\infty$, which is a geometric limit of the surfaces $\tilde{g}_i$, is incompressible, hence $\tilde{g}_\infty(c)$ is not null-homotopic.)

First consider the case (i). Since $i(\lambda, c) = 0$, either there exists a component λ_1 of λ such that $\lambda_1 \cap c = \emptyset$ or the support of λ is equal to c. In the latter case, we can retake c such that $\lambda \cap c = \emptyset$, and reduce this case to the former. Hence we can assume that the existence of such λ_1. For λ_1 and $\tilde{g}_\infty$ as above, the alternative (2) of Lemma 6.10 cannot hold true. If it did, we could not find a map h_∞ homotopic to $\tilde{g}_\infty$ such that length$(h_\infty(\lambda_1))$ goes to 0. On the other hand, the alternative (1) in Lemma 6.10 holds true for λ by Proposition 6.11, and since λ_1 is a sublamination of λ, the same would hold for λ_1. This is a contradiction. Thus, the condition of the alternative (1) in Lemma 6.10 must be valid for λ_1.

Let μ be a measured lamination obtained by adding c with weight 1 to λ_1 as a compact leaf. Then the alternative (1) of Lemma 6.10 cannot hold true for μ because if it did, the alternative (1) would hold also for c, while $\tilde{g}_\infty(c)$ is homotopic to a closed geodesic as we assumed above. Therefore by Lemma 6.10, the condition of the alternative (2) must be satisfied. Then by the same argument as in the proof of Proposition 5.1 in [5], we can homotope h_∞ so as to make the total curvature and the quadratic variation of angle of $h_\infty(\mu)$ arbitrarily small keeping its image in a compact set, where h_∞ is adapted to a tied neighbourhood of a train track carrying μ. This contradicts the fact that for λ_1, the alternative (1) is valid, which was proved above.

Thus the remaining possibility is the case (ii) when $\tilde{g}_\infty(c)$ represents a parabolic element. In this case, there exist closed curves $\{c_k\}$ in $\tilde{M}_\infty$ freely homotopic to $\tilde{g}_\infty(c)$ whose lengths and geodesic curvatures go to 0. Since the curvature of $\tilde{M}_\infty$ is pinched, this is possible only when c_k goes away from any compact set as $k \to \infty$ by Margulis' lemma for negatively curved manifolds. In particular, for sufficiently large k, the c_k cannot intersect the neighbourhood $\tilde{V}_\infty$ of the preimage of the branching locus, which implies that the c_k are contained in the region where the branched covering $p_\infty : \tilde{M}_\infty \to \mathbf{H}^3/\Gamma'$ is locally isometric. Hence the length and the geodesic curvature of $p_\infty(c_k)$ go to 0 as $k \to \infty$. Then the closed curve $p_\infty \circ \tilde{g}_\infty(c)$ also represents a parabolic element because there exists a closed curve freely homotopic to it with arbitrarily small length and geodesic curvature. This contradicts the assumption that Γ' has no parabolic elements. □

We shall see that by virtue of Proposition 6.11, we can get information on the end of $\mathbf{H}^3/\Gamma'$ facing S. As a first step, we shall show that for simple closed curves on S whose projective classes converge to a projective lamination with the same support as $[\lambda]$, the closed geodesics homotopic to them tend to an end of $\mathbf{H}^3/\Gamma'$ after taking a subsequence.

LEMMA 6.14. *Let λ' be a measured lamination on S with the same support as the λ defined before to whose projective class the hyperbolic structures $\{m_i\}$ converge. Let $\{\delta_j\}$ be a sequence of simple closed curves on S such that $\{\delta_j/\mathrm{length}_S(\delta_j)\}$ converges to $\lambda'/\mathrm{length}_S(\lambda')$ in $\mathcal{ML}(S)$, where* length$_S$ *denotes the geodesic length with respect to a fixed hyperbolic structure on S. Then for any compact set K in $\tilde{M}_\infty$, there exists j_0 such that if $j \geq j_0$, then the closed geodesic δ_j^* freely homotopic to $\tilde{g}_\infty(\delta_j) \subset \tilde{M}_\infty$ is disjoint from K.*

PROOF. Suppose, seeking a contradiction, that there exists a compact set K intersecting infinitely many δ_j^*. For each δ_j^*, we can construct a simplicial ruled surface $d_j : S \to \tilde{M}_\infty$ with respect to a triangulation containing δ_j on the union of an edge and a vertex, which is homotopic to $\tilde{g}_\infty$ such that $d_j(\delta_j) = \delta_j^*$. By Theorem 3.2.7 in Canary [**8**], there is a uniform upper bound $Q > 0$ for the diameters of the $d_j(S)$'s. Therefore there exists a compact set N_∞ containing all the images of d_j. We can assume that N_∞ is a connected irreducible 3-manifold by enlarging it if necessary. Furthermore, since the d_j's are incompressible, we can make the boundary of N_∞ incompressible by performing surgery not touching the images of d_j. We can also assume that no two boundary components of N_∞ are homotopic in $\tilde{M}_\infty - \mathrm{Int}N_\infty$. In this situation, by elementary 3-dimensional topology, we can see that each two of the d_j's are homotopic in N_∞ as they are homotopic in $\tilde{M}_\infty$. In particular, δ_j^* is homotopic to $d_1(\delta_j)$ in N_∞.

Since N_∞ is compact, by the main theorem of Milnor [**32**], there exists a constant $L > 0$ such that

$$L^{-1}\mathrm{wl}(\delta_j) \leq \mathrm{length}(\delta_j^*) \leq L\mathrm{wl}(\delta_i),$$

where the wl denotes the minimum word length of the elements of the conjugacy class of δ_j with respect to a fixed generator system of $\pi_1(S)$. Furthermore, $\mathrm{wl}(\delta_j) = O(\mathrm{length}_S(\delta_j))$ by the same theorem. Therefore, we have a constant $L' > 0$ such that

$$L'^{-1}\mathrm{length}_S(\delta_j) \leq \mathrm{length}(\delta_j^*) \leq L'\mathrm{length}_S(\delta_j).$$

On the other hand, by Proposition 6.11, for any $\epsilon > 0$, there exists a continuous map $h_\epsilon : S \to \tilde{M}_\infty$ homotopic to $\tilde{g}_\infty$, which is adapted to a tied neighbourhood of a train track τ_ϵ carrying λ' such that $\mathrm{length}(h_\epsilon(\lambda')) \leq \epsilon$. For each τ_ϵ, there exists an integer j_1 such that if $j > j_1$ then δ_j is carried by τ_ϵ; for, since λ' is maximal and connected, the train track τ_ϵ must correspond to a coordinate neighbourhood around λ'. Since $\{\delta_j/\mathrm{length}_S(\delta_j)\}$ converges to $\lambda'/\mathrm{length}_S(\lambda')$, for any $\epsilon > 0$, there exists an integer j_2 such that if $j > j_2$ then $h_\epsilon(\delta_j)$ has length less than $2\epsilon\mathrm{length}_S(\delta_j)/\mathrm{length}_S(\lambda')$. Thus we have $\mathrm{length}(\delta_j^*)/\mathrm{length}_S(\delta_j) \to 0$ as $j \to 0$, and this contradicts the inequality proved above. □

Let $\overline{\delta}_j^*$ be the closed geodesic in $\mathbf{H}^3/\Gamma'$ freely homotopic to $\delta_j \simeq p_\infty(\delta_j^*)$. (Recall that $g_\infty = p_\infty \circ \tilde{g}_\infty$ is homotopic to the inclusion.) Since δ_j^* does not intersect $\tilde{V}_\infty$ for sufficiently large j, we have $\overline{\delta}_j^* = p_\infty(\delta_j^*)$ for large j. Since p_∞ is finite-sheeted, for any compact set K in $\mathbf{H}^3/\Gamma'$, there exists an integer j_0 such that $\overline{\delta}_j^* \cap K = \emptyset$ if $j > j_0$.

LEMMA 6.15. *The measured lamination λ is contained in the Masur domain when S is regarded as the exterior boundary of the compact core C'.*

PROOF. Since we have already shown that λ is connected and maximal, the only possibility that λ is not contained in the Masur domain is that there exists a measured lamination λ' in $\overline{\mathcal{C}}(S)$ with the same support as λ. If so, then there exists a sequence of simple closed curves $\{b_k\}$ on S bounding compressing discs in C' such that $[b_k] \to [\lambda']$ in $\mathcal{PL}(S)$. Since b_k represents the trivial element in Γ', so does $p_\infty \circ \tilde{g}_\infty(b_k) = g_\infty(b_k) \simeq b_k$. It follows that the closed geodesic b_k^* homotopic to $\tilde{g}_\infty(b_k)$ cannot be disjoint from the neighbourhood $\tilde{V}_\infty$ of the preimage of the branching locus because otherwise $p_\infty(b_k^*)$ would be a null-homotopic closed geodesic. Thus b_k^* intersects the compact set $\tilde{V}_\infty$ as $k \to \infty$. This cannot happen, however, by Lemma 6.14. □

LEMMA 6.16. *The closed geodesic $\overline{\delta}_j^*$ tends to the end facing S as $j \to \infty$.*

PROOF. By the remark just before Lemma 6.15, the closed geodesic $\overline{\delta}_j^*$ cannot stay in any given compact set as $j \to \infty$. Let e be the end of $\mathbf{H}^3/\Gamma'$ facing the boundary component S of the compact core C'. What remains to prove is that no subsequence of $\{\overline{\delta}_j^*\}$ can go to an end other than e.

Suppose, seeking a contradiction, that a subsequence of $\{\overline{\delta}_j^*\}$, which we shall denote by the same symbol, goes to an end other than e. Then for sufficiently large j, the closed geodesic $\overline{\delta}_j^*$ is contained in a component E' of the complement of the compact core C' which faces a boundary component S' of C' other than S. Now, since C' is a compact core, there exists a closed curve δ_k' in C' which is freely homotopic to $\overline{\delta}_j^*$. Let $A : S^1 \times I \to \mathbf{H}^3/\Gamma'$ be a homotopy from $\overline{\delta}_j^*$ to δ_j', which we take to be transverse to S'. As S' is separating, there exists an essential simple closed curve s on $S^1 \times I$ such that $A(s) \subset S'$. This implies that δ_j is freely homotopic to a closed curve $A(s)$ on S'. Since C' is a compact core, we can take a homotopy between δ_j and $A(s)$ in C'.

Since S' is a component of the interior boundary of C', for any closed curve δ on S' there exists a compression disc of C' disjoint from δ. As is shown in Lemmata 5.4, 5.5 in Canary [**7**], no simple closed curve in the Masur domain $\mathcal{M}(S)$ can be homotopic to such a closed curve. Since we know, by Lemma 6.15, that δ_j is contained in the Masur domain for sufficiently large j, and by the argument above, that δ_j is homotopic to a closed curve on S', this is a contradiction. □

Thus the sequence $\{\delta_j\}$ above is exactly what we wanted in Theorem 6.1, and we have completed the proof of Theorem 6.1.

CHAPTER 7

Strong convergence of function groups

Recall that for a Kleinian group G as in Theorem 2.1, we took a subgroup Γ corresponding to $\Psi_{\#}^{-1}\iota_{\#}\pi_1(S)$. We have quasi-conformal deformations (Γ_i, ϕ_i) of Γ and its algebraic limit (Γ', ϕ) which are subgroups of quasi-conformal deformations (G_i, ψ_i) and its algebraic limit (G', ψ). We want to apply Theorem 6.1 to these $\Gamma, (\Gamma_i, \phi_i)$, and (Γ', ϕ). To show that the theorem is applicable in our situation, we have only to prove that $\mathbf{H}^3/\Gamma$ has a compact core which is a compression body, and that the quasi-conformal deformations Γ_i converge to Γ' strongly. We defer the proof of the former to the next chapter. We shall prove in this section that the latter is true if Γ' has non-empty region of discontinuity and $\mathbf{H}^3/\Gamma$ has a compact core which is homeomorphic to a compression body. If Γ' has empty region of discontinuity, then so does G', and the conclusion of Theorem 2.1 follows trivially. Thus it is sufficient to prove Theorem 2.1 under the assumption that Γ' has non-empty region of discontinuity.

THEOREM 7.1. *Let Γ be a non-free geometrically finite Kleinian group without parabolic elements such that $\mathbf{H}^3/\Gamma$ has a compact core C which is a compression body. Let $\{(\Gamma_i, \phi_i)\}$ be a sequence of quasi-conformal deformations of Γ with isomorphisms ϕ_i, which converges algebraically to a Kleinian group Γ' without parabolic elements with an isomorphism $\phi : \Gamma \to \Gamma'$. In this situation,*

(1) *The hyperbolic 3-manifold $\mathbf{H}^3/\Gamma'$ has a compact core C' which is a compression body such that the homotopy equivalence $\Phi : \mathbf{H}^3/\Gamma \to \mathbf{H}^3/\Gamma'$ induces a homeomorphism from C to C'. (In other words, the homotopy equivalence Φ can be homotoped so that $\Phi|C$ is a homeomorphism to C'.)*
(2) *Furthermore if $\Omega_{\Gamma'} \neq \emptyset$ and the end facing $\partial_e C'$ is geometrically infinite, then $\{(\Gamma_i, \phi_i)\}$ converges to (Γ', ϕ) strongly, that is, $\{\Gamma_i\}$ converges to Γ' also geometrically.*

To prove Theorem 7.1-(2), we have only to prove that every subsequence of $\{\Gamma_i\}$ in the statement has a subsequence which converges geometrically to Γ'. Since any subsequence of $\{\Gamma_i\}$ has a subsequence which converges geometrically to a Kleinian group containing Γ', we can assume that $\{\Gamma_i\}$ converges geometrically to a Kleinian group Γ_∞ which contains Γ'. Then what we have to prove is that $\Gamma_\infty = \Gamma'$.

We can assume that C is the convex core of $\mathbf{H}^3/\Gamma$ because Γ is geometrically finite and a compact core is unique up to homeomorphism by McCullogh-Miller-Swarup [**30**] as cited in §1.B. Let T be an incompressible boundary component of C. Let C_i be the convex core of $\mathbf{H}^3/\Gamma_i$, and let T_i be the boundary component of C_i homotopic to $\Phi_i(T)$, which is uniquely determined. Then there is a homeomorphism $h_i' : T \to T_i$ homotopic to $\Phi_i|T$, which is unique up to homotopy in $\mathbf{H}^3/\Gamma_i$ because T is incompressible and $\mathbf{H}^3/\Gamma_i$ is not a surface bundle over a circle. (Refer to the proof of Theorem 5.2.18 in Canary-Epstein-Green [**10**].) Let n_i be the element of the Teichmüller space $\mathcal{T}(T)$ represented by the marked hyperbolic structure corresponding to the pair of the hyperbolic structure on T_i induced from the metric on $\mathbf{H}^3/\Gamma_i$, and the homeomorphism h_i'. Then $\{n_i\}$ has a subsequence which converges either in $\mathcal{T}(T)$ or to a projective lamination in the Thurston compactification of $\mathcal{T}(T)$. We shall first consider the case when $\{n_i\}$ has a convergent subsequence in the Teichmüller space.

LEMMA 7.2. *Suppose that $\{n_i\}$ has a subsequence which converges in $\mathcal{T}(T)$. Then $\{T_i\}$ converges geometrically, after taking a subsequence, to a boundary component T_∞ of the convex core of $\mathbf{H}^3/\Gamma_\infty$, which can be lifted to a boundary component T' of the convex core of $\mathbf{H}^3/\Gamma'$ homotopic to $\Phi(T)$.*

PROOF. The proof of this lemma is entirely the same as a part of the proof of Theorem 2.1 in Ohshika [**35**] (pp. 102-103). □

Next let us consider the case when $\{n_i\}$ converges, after taking a subsequence, to a projective lamination in the Thurston compactification of the Teichmüller space.

LEMMA 7.3. *Suppose that $\{n_i\}$ has a subsequence which converges to a projective lamination $[\lambda]$ in the Thurston compactification of $\mathcal{T}(T)$. Then $\mathbf{H}^3/\Gamma'$ has a simply degenerate end e facing the boundary component of the compact core C' of $\mathbf{H}^3/\Gamma'$ that is homotopic to $\Phi(T)$.*

PROOF. We shall first prove that λ cannot be realized by a pleated surface homotopic to $\Phi|T$ by contradiction. (Although this can also be proved by using the continuity of the length function as in [**35**] and [**37**], we shall use the lemmata in the previous sections instead.)

Suppose, on the contrary, that λ can be realized by a pleated surface $f' : T \to \mathbf{H}^3/\Gamma'$ homotopic to $\Phi|T$. Consider a subgroup Γ^T of Γ associated with $\pi_1(T)$. Let Γ_i^T denote $\phi_i(\Gamma^T)$, and let Γ_T' denote $\phi(\Gamma^T)$. By lifting f' to $\mathbf{H}^3/\Gamma_T'$, we can see that the measured lamination λ can be realized by a pleated surface homotopic to a map $\widetilde{\Phi|T} : T \to \mathbf{H}^3/\Gamma_T'$ which is a lift of $\Phi|T$. Then by Lemma 4.10, for any sequence of weighted simple closed curves $\{w_k c_k\}$ converging to λ, and given $t < 1, \delta > 0$, there exist a subsequence $\{c_{k(l)}\}$ of $\{c_k\}$, a train track τ carrying both λ and the $c_{k(l)}$'s, and a map $h : T \to \mathbf{H}^3/\Gamma_T'$ homotopic to f' as follows. The map h is adapted to a tied neighbourhood N_τ of τ, and for sufficiently large l, the closed geodesic $c_{k(l)}^*$

homotopic to $h(c_{k(l)})$ in $\mathbf{H}^3/\Gamma'_T$ has a part with length $t\text{length}h(c_{k(l)})$ which is contained in the δ-neighbourhood of $h(c_{k(l)})$.

By the same argument as the proof of Lemma 6.12, for any $t < 1$ and $\delta > 0$ there exist integers i_0, l_0 and maps $h_i : T \to \mathbf{H}^3/\Gamma_i^T$ homotopic to the lifts $\widetilde{\Phi_i|T}$ of $\Phi_i|T$ such that if $i > i_0, l > l_0$, the closed geodesic $c^*_{i,k(l)}$ homotopic to $h_i(c_{k(l)})$ in $\mathbf{H}^3/\Gamma_i^T$ has a part of length at least $t\text{length}h_i(c_{k(l)})$ lying in the δ-neighbourhood of $h_i(c_{k(l)})$, and such that $\text{length}(h_i(\lambda))$ does not go to 0 as $i \to \infty$. This implies that $\{w_i\text{length}(c^*_{i,k(l)})\}$ does not go to 0 as $i, l \to \infty$.

On the other hand, we can take a sequence $\{w_i c_i\}$ converging to λ such that $w_i\text{length}_{(S,n_i)}(c_i)$ goes to 0 as $i \to 0$ as is shown in Theorem 2.2 of Thurston [**49**]. Then we have $w_i\text{length}(c^*_{i,i}) \to 0$ by Sullivan's theorem proved in Epstein-Marden [**15**] in which a universal constant K such that $\text{length}(c^*_{i,i}) \leq K\text{length}_{(S,n_i)}(c_i)$ is given. Applying the result in the last paragraph to this $\{c_i\}$, we get a contradiction. Thus we have proved that λ cannot be realized by a pleated surface homotopic to $\Phi|T$.

Since T is incompressible, by the argument of pp.88-89 of Bonahon [**5**] or §9.4.1 of Thurston [**45**], we can see that there exists a simply degenerate end of $\mathbf{H}^3/\Gamma'$ which has a neighbourhood parametrized by $T' \times \mathbf{R}$ where $T' \times \{t\}$ is homotopic to $\Phi(T)$. Then we can construct a compact core of $\mathbf{H}^3/\Gamma'$ one of whose boundary component is equal to the surface corresponding to $T' \times \{0\}$. Since the isotopy classes of the incompressible boundary components of a compact core do not depend on the choice of a compact core ([**30**]), it follows that C' has a boundary component homotopic to $\Phi|T$ which faces the simply degenerate end. This completes the proof. □

PROOF OF THEOREM 7.1-(1). Since Γ is assumed not to be free, the compact core C is not a handlebody. Let $T_1, \ldots, T_m$ be the components of the interior boundary $\partial_i C$. By Lemmata 7.2 and 7.3, $\Phi|T_k$ is homotopic to either a homeomorphism onto a boundary component T'_k of the convex core of $\mathbf{H}^3/\Gamma'$ or an embedding with its image T'_k contained in a neighbourhood of simply degenerate end, which is homeomorphic to $T'_k \times \mathbf{R}$. If $k \neq k'$, then T'_k and $T'_{k'}$ are not homotopic in $\mathbf{H}^3/\Gamma'$. Therefore by taking embeddings far enough in neighbourhoods of simply degenerate ends, we can assume that T'_k and $T'_{k'}$ are disjoint if $k \neq k'$. Furthermore since T'_k is either a boundary component of the convex core or facing a simply degenerate end homeomorphic to $T'_k \times \mathbf{R}$, in either case, the surface T'_k cuts off a neighbourhood E_k of an end, which is homeomorphic to $T'_k \times \mathbf{R}$, from $\mathbf{H}^3/\Gamma'$.

Now, since C is a compression body, there is a deformation retract of C consisting of $\sqcup_{k=1}^m T_k$ and properly embedded arcs $\alpha_1, \ldots, \alpha_n$ all of whose endpoints lie on $\sqcup_{k=1}^m T_k$, which has a regular neighbourhood isotopic to C. We homotope Φ so that $\Phi|T_k$ is a homeomorphism to T'_k. Since T'_k cuts off $E_k \cong T'_k \times \mathbf{R}$ from $\mathbf{H}^3/\Gamma'$, we can further homotope Φ so that the arcs $\Phi(\alpha_j)$

are mutually disjoint, and so that $\Phi(\alpha_j)$ does not intersect the components E_k. Since after moving by such a homotopy, all the arcs $\Phi(\alpha_j)$ are contained in $\mathbf{H}^3/\Gamma' - \cup_{k=1}^m E_k$, which is one of the components of the complement of $\sqcup_{k=1}^m T'_k$, a regular neighbourhood of $\sqcup_{k=1}^m T'_k \cup (\cup_{l=1}^n \Phi(\alpha_l))$ is a compression body, which we shall denote C'. It is easy to see that C' is a compact core of $\mathbf{H}^3/\Gamma'$ and that $\Phi|C$ is homotopic to a homeomorphism to C' using the assumptions that C is a compact core of $\mathbf{H}^3/\Gamma$ and that Φ is a homotopy equivalence. □

Proof of Theorem 7.1-(2). By assumption, at least one of the ends of $\mathbf{H}^3/\Gamma'$ facing $T'_1, \ldots, T'_m$ is geometrically finite. By renumbering $T_1, \ldots, T_m$, we can assume that the ends facing $T'_1, \ldots, T'_\mu$ are geometrically finite and those facing $T'_{\mu+1}, \ldots, T'_m$ are geometrically infinite. Let $\Sigma_1, \ldots, \Sigma_\mu$ be the components of $\Omega_{\Gamma'}/\Gamma'$ corresponding to $T'_1, \ldots, T'_\mu$.

Let $\pi : \mathbf{H}^3/\Gamma' \to \mathbf{H}^3/\Gamma_\infty$ be the covering projection associated with the inclusion $\Gamma' \subset \Gamma_\infty$. By Lemma 7.2, $\pi|T'_1 \cup \ldots \cup T'_\mu$ is a homeomorphism to a union of boundary components of the convex core of $\mathbf{H}^3/\Gamma_\infty$. Therefore $\Sigma_1, \ldots, \Sigma_\mu$ can be regarded also as components of $\Omega_{\Gamma_\infty}/\Gamma_\infty$. For $k = 1, \ldots, \mu$, let Ω_k be a component of $\Omega_{\Gamma'}$ which is projected to Σ_k in Ω'_Γ/Γ'. Since Σ_k is a component of $\Omega_{\Gamma_\infty}/\Gamma_\infty$, the domain Ω_k is also a component of Ω_{Γ_∞}. Furthermore since the end facing $\partial_e C'$ is geometrically infinite by assumption, the boundary of the convex core of $\mathbf{H}^3/\Gamma'$ is equal to $T'_1 \cup \ldots \cup T'_\mu$, which implies that $\Omega_{\Gamma'}/\Gamma' = \Sigma_1 \cup \cdots \cup \Sigma_\mu$.

Suppose, seeking a contradiction, that $\Gamma_\infty \neq \Gamma'$. Let γ be an element in $\Gamma_\infty - \Gamma'$. Consider the component $\gamma\Omega_1$ of Ω_{Γ_∞}. Because $\Gamma' \subset \Gamma_\infty$, we have $\Omega_{\Gamma_\infty} \subset \Omega_{\Gamma'}$. Since $\Omega_{\Gamma'}/\Gamma' = \Sigma_1 \cup \cdots \cup \Sigma_\mu$, we have $\Omega_{\Gamma'} = \cup_{g \in \Gamma'} g(\Omega_1 \cup \ldots \cup \Omega_\mu)$, and for any $g \in \Gamma'$ and k, the domain $g\Omega_k$ is a component of both $\Omega_{\Gamma'}$ and Ω_{Γ_∞}. Therefore there exist $g \in \Gamma'$ and $\Omega_k (1 \leq k \leq \mu)$ such that $\gamma\Omega_1 \subset g\Omega_k$. As $g\Omega_k$ is also a component of Ω_{Γ_∞}, and $g\Omega_k$ is projected to the component Σ_k of $\Omega_{\Gamma_\infty}/\Gamma_\infty$, this implies that $\gamma\Omega_1 = g\Omega_k$, that $k = 1$, and that $g^{-1}\gamma$ is contained in the stabilizer of Ω_1 in Γ_∞. On the other hand, since $\Omega_{\Gamma_\infty}/\Gamma_\infty$ contains Σ_1 as a component, the stabilizer of Ω_1 in Γ_∞ is equal to that in Γ'. Hence $g^{-1}\gamma$ must be contained in Γ'. This contradicts the choice of γ. □

CHAPTER 8

Proof of the main theorem

In this section, we shall finally prove our main theorem Theorem 2.1. Before starting a proof in general, we prove Theorem 2.1 in a special case when $\mathbf{H}^3/G$ has a compact core that is a compression body, which would be also useful to illustrate a general line of the argument.

8.A. A special case

Suppose that $C(G)$ is a compact core of $\mathbf{H}^3/G$, which we assume to be a compression body. Then the subgroup Γ of G, which is associated to the image of the fundamental group of $\partial_e C(G)$ in $\pi_1(C(G))$, is equal to G. We shall use the symbols Γ, Γ_i and Γ', instead of G, G_i and G' for denoting the Kleinian groups in question; a geometrically finite group, its quasi-conformal deformations, and their algebraic limit in Theorem 2.1, in order to emphasize that we are dealing with the special case. Also, we denote a compact core of $\mathbf{H}^3/\Gamma$ by C in this special case.

By Theorem 7.1-(1), there exists a compact core C' of $\mathbf{H}^3/\Gamma'$ and the map $\Phi|C$ is homotopic to a homeomorphism to C'. Suppose first that the end facing $\partial_e C'$ is geometrically finite. Then as the boundary components of C' other than $\partial_e C'$ are incompressible, all the ends of $\mathbf{H}^3/\Gamma'$ satisfy the assumption of Theorem 4.1 by Bonahon's theorem in [**5**]. Thus by using Theorem 4.1 when G is not a free group, and by Maskit's theorem when G is a free group, we can prove Theorem 2.1.

Next suppose that the end facing $\partial_e C'$ is geometrically infinite. Then either we have $\Omega_{\Gamma'} = \emptyset$ or the assertion of Theorem 7.1-(2) holds true. If $\Omega_{\Gamma'} = \emptyset$, the conclusion of Theorem 2.1 holds trivially. Suppose that the assertion of Theorem 7.1-(2) holds. Then as $\{(\Gamma_i, \phi_i)\}$ converges to (Γ, ϕ) strongly, the assumption of Theorem 6.1 is satisfied, which implies that on $S = \partial_e C'$, there exists a sequence of simple closed curves $\{\delta_j\}$ whose projective classes converge inside $\mathcal{PM}(S)$ such that the closed geodesics δ_j^* freely homotopic to δ_j in $\mathbf{H}^3/\Gamma'$ tend to the end facing S.

Again if Γ' is a free group, Theorem 2.1 holds true by Maskit's theorem [**26**] stating that purely loxodromic geometrically infinite free Kleinian groups have empty region of discontinuity. Therefore we can assume that Γ' is not a free group. Since the ends of $\mathbf{H}^3/\Gamma'$, other than the one facing S, face incompressible components of ∂C, for these ends, the assumption of Theorem 4.1 is satisfied by Bonahon's theorem as before. This fact together with the conclusion of Theorem 6.1 above makes it possible to apply Theorem 4.1 to our Γ', which implies the conclusion of Theorem 2.1 for Γ'.

Thus we have completed the proof of our main theorem in the special case when a compact core of $\mathbf{H}^3/G$ is a compression body.

8.B. The existence of a homeomorphism

Now let us consider the general case. To apply the result for the special case above to the general case, we have to prove that for a compact core $C(G)$ of $\mathbf{H}^3/G$, the map $\Psi|C(G)$ is homotopic to a homeomorphism to a compact core of $\mathbf{H}^3/G'$. We state this as a proposition here.

PROPOSITION 8.1. *Let $C(G)$ be a compact core of $\mathbf{H}^3/G$. Then there exists a compact core $C(G')$ of $\mathbf{H}^3/G'$ such that $\Psi|C(G)$ is homotopic to a homeomorphism to $C(G')$.*

Let $V(G)$ be the characteristic compression body of $C(G)$. Let $N(G)$ be $\overline{C(G) - V(G)}$, which is an irreducible and boundary-irreducible compact 3-manifold. To prove Proposition 8.1, we shall first show that $\Psi|N(G)$ can be homotoped to an embedding such that its projection to the geometric limit $\mathbf{H}^3/G_\infty$ of $\{\mathbf{H}^3/G_i\}$ is still an embedding. (Lemma 8.2 below.) After that, regarding $C(G)$ as the union of $N(G)$ and one-handles, we shall prove that the images of one-handles by Ψ can be homotoped off $\mathrm{Int}N(G)$ to complete the proof of the proposition. See Figure 1 illustrating schematically the situation which we should have in mind.

Now let us state the key lemma which shows that Ψ can be homotoped so that for the boundary-irreducible part $N(G)$, the restriction $\Psi|N(G)$ and its projection to the geometric limit are embeddings. Note that, as before, by taking a subsequence, we can assume that $\{G_i\}$ converges geometrically to a Kleinian group containing G'. We use the same symbol $\{G_i\}$ to denote such a geometrically convergent sequence.

LEMMA 8.2. *Let G_∞ be the geometric limit of $\{G_i\}$ and let $p_\infty : \mathbf{H}^3/G' \to \mathbf{H}^3/G_\infty$ be the covering associated with the inclusion $G' \subset G_\infty$. Then Ψ can be homotoped so that both $\Psi|N(G) : N(G) \to \mathbf{H}^3/G'$ and $p_\infty \circ \Psi|N(G) : N(G) \to \mathbf{H}^3/G_\infty$ are embeddings.*

8.C. Lemmata for the proof of Lemma 8.2

We need several steps to prove Lemma 8.2.

Let N_0 be a component of $N(G)$ which is not a product I-bundle over a closed (orientable) surface, and let $G(N_0)$ be a subgroup of G corresponding to $\pi_1(N_0)$. Let $G_i(N_0)$ be $\psi_i(G(N_0))$, and let $G'(N_0)$ be $\psi(G(N_0))$.

LEMMA 8.3. *The sequence $\{(G_i(N_0), \psi_i|G(N_0))\}$ converges strongly to $(G'(N_0), \psi|G(N_0))$.*

PROOF. Since $\mathbf{H}^3/G(N_0)$ has a boundary-irreducible compact core, which is homeomorphic to N_0, and $G'(N_0)$ has no parabolic elements, this lemma is a corollary of Theorem 2.1 in Ohshika [**35**], which is originally due to Thurston [**45**]. (Note that the exclusion of Kleinian groups isomorphic to surface groups in Theorem 2.1 in [**35**] is necessary only for those isomorphic to closed orientable surface groups.) □

LEMMA 8.4. *Let $C(N_0)$ be a compact core of $\mathbf{H}^3/G(N_0)$ and let $C'(N_0)$ be that of $\mathbf{H}^3/G'(N_0)$. Let $\Psi^{N_0} : \mathbf{H}^3/G(N_0) \to \mathbf{H}^3/G'(N_0)$ be a homotopy equivalence corresponding to $\psi|G(N_0)$. Then $\Psi^{N_0}|C(N_0)$ is homotopic to a homeomorphism to $C'(N_0)$. Furthermore we can choose a homotopy which fixes the complement of an arbitrarily small neighbourhood of $C(N_0)$.*

PROOF. This was already proved in the proof of Theorem 2.1 in Ohshika [**35**]. (pp. 104-105) □

DEFINITION 8.5. Let N_k be a component of $N(G)$. We divide the boundary components of N_k into two families: the one consisting of the components that are also contained in $\partial C(G)$, which we call the *external boundary components*; and the other consisting of the components that are not contained in $\partial C(G)$, in other words, those contained in $\partial V(G)$, which we call the *internal boundary components.* We use these terms also for boundary components of a boundary-irreducible part $N(G')$ of a compact core $C(G')$ of $\mathbf{H}^3/G'$. Do not confuse them with the interior or the exterior boundary of a compression body.

For a boundary component Σ of N_0, let $G(\Sigma)$ be a subgroup of $G(N_0)$ corresponding to $\pi_1(\Sigma)$, and let $G_i(\Sigma)$ and $G'(\Sigma)$ be $\psi_i(G(\Sigma))$ and $\psi(G(\Sigma))$ respectively.

LEMMA 8.6. *The Kleinian group $G'(\Sigma)$ is geometrically finite, hence quasi-Fuchsian for every internal boundary component Σ of N_0.*

PROOF. We shall prove this lemma by contradiction. Suppose, seeking a contradiction, that $G'(\Sigma)$ is geometrically infinite for some internal boundary component Σ of N_0. Since $G'(\Sigma)$ is isomorphic to the fundamental group of a closed surface and has no parabolic elements, by Bonahon's theorem [**5**], there is an end of $\mathbf{H}^3/G'(\Sigma)$ which is simply degenerate. On the other hand, there exists a component V of $V(G)$ such that the interior boundary $\partial_i V$ contains Σ as a component, by the definition of internal boundary component. Let W be the submanifold $N_0 \cup V$, and let $G(W)$ be a subgroup of G corresponding to the image of the fundamental group of W, which contains $G(N_0)$. Then W can be lifted homeomorphically to a compact core $\tilde{W}$ of $\mathbf{H}^3/G(W)$, and N_0 is also lifted homeomorphically to a submanifold $\tilde{C}(N_0)$ of $\tilde{W}$. The surface Σ is lifted homeomorphically to a boundary component $\tilde{\Sigma}$ of $\tilde{C}(N_0)$. Let $G'(W)$ be $\psi(G(W))$ and let $\psi^W : G(W) \to G'(W)$ be the restriction of ψ. Let $p'_W : \mathbf{H}^3/G'(\Sigma) \to \mathbf{H}^3/G'(W)$ be the covering projection associated with the inclusion $G'(\Sigma) \subset G'(W)$.

Let e be a simply degenerate end of $\mathbf{H}^3/G'(\Sigma)$. By Thurston's covering theorem in [**45**] (see also Lemma 2.2 in Ohshika [**35**] and Canary [**9**]), there exists a neighbourhood E of the end e such that $p'_W|E$ is proper and a finite-sheeted covering to its image. In particular, $\mathbf{H}^3/G'(W)$ has an end which has a neighbourhood parametrized homeomorphically by $\Xi \times \mathbf{R}$ for a closed orientable surface Ξ such that $\Psi^W|\tilde{\Sigma}$ is homotopic to a covering map onto $\Xi \times \{t\}$.

If $\Psi^W|\tilde{\Sigma}$ is homotopic to a non-trivial covering to $\Xi \times \{t\}$, then there exists a (possibly singular) closed incompressible orientable surface Ξ_0 in $\mathbf{H}^3/G$ such that the inclusion of Σ can be homotoped to a non-trivial covering onto Ξ_0, which can be obtained by pulling back the projected image of $\Xi \times \{t\}$ in $\mathbf{H}^3/G'$ by Ψ^{-1}. This is impossible because Σ is an embedded incompressible surface and Ξ_0 is orientable. (This can be seen, for instance, as follows. Consider a covering of $\mathbf{H}^3/G$ associated with the fundamental group of Ξ_0, which is homeomorphic to $\Xi_0 \times \mathbf{R}$ because G is geometrically finite. Then Σ can be lifted homeomorphically to an embedded surface in the covering, whose fundamental group is a proper subgroup of $\pi_1(\Xi_0 \times \{t\})$. This is impossible.) Thus $\Psi^W|\tilde{\Sigma}$ must homotopic to a homeomorphism onto $\Xi \times \{t\}$.

Let $C'(W)$ be a compact core of $\mathbf{H}^3/G'(W)$ one of whose boundary component is $\Xi \times \{t\}$ with respect to the parametrization of the neighbourhood of the end by $\Xi \times \mathbf{R}$. (Such a compact core exists by the relative core theorem due to McCullough [**28**].) Since $C'(W)$ is a compact core, we can homotope Ψ^W so that its image is contained in $C'(W)$. By the same argument as Lemmata 7.2 and 7.3, the restriction of Ψ^W to each component of $\partial\tilde{C}(N_0) - \tilde{\Sigma}$ is homotopic to a homeomorphism to a component of $\partial C'(W)$ because $\partial\tilde{C}(N_0) - \tilde{\Sigma}$ is contained in $\partial\tilde{W}$, and a homotopy can be taken to be contained in $C'(W)$. Hence $\Psi^W|\tilde{C}(N_0)$ is homotopic to a continuous map $\hat{\Psi} : \tilde{C}(N_0) \to C'(W)$ such that $\hat{\Psi}|\partial\tilde{C}(N_0)$ is a homeomorphism into $\partial C'(W)$ because of the remarks above and the fact that no two components of $\tilde{C}(N_0)$ are mutually homotopic.

Then by Theorem 13.6 in Hempel [**22**], (which is a generalization of Waldhausen's theorem in [**52**],) $\hat{\Psi}$ is homotopic to a covering map fixing $\hat{\Psi}|\partial\tilde{C}(N_0)$ because we assumed that N_0 is not a product I-bundle, and in the case when N_0 is a twisted I-bundle, $\hat{\Psi}(\tilde{C}(N_0))$ cannot be homotopic into $\partial C'(W)$ which is orientable. (Recall that maps homotopic into a boundary component of $\partial C'(W)$ are the only exceptions for which Hempel's Theorem 13.6 does not give us covering maps by homotoping maps as above.) We can also see that $\hat{\Psi}$ is homotopic to a homeomorphism again by Hempel's Theorem 13.6 because $\hat{\Psi}|\partial\tilde{C}(N_0)$ is a homeomorphism. This contradicts the fact that $\pi_1(N_0)$ is a proper subgroup of $\pi_1(W)$. □

We need to prove a similar lemma for a component N_p of $N(G)$ which is homeomorphic to a product I-bundle over a closed orientable surface, neither of whose boundary components is contained in $\partial C(G)$. This will not be used until the last stage of the proof of Proposition 8.1.

LEMMA 8.7. *Let N_p be a component of $N(G)$ which is homeomorphic to a product I-bundle over a closed orientable surface. Suppose that both of the components of ∂N_p are interior boundary components. Let $G(N_p)$ be a subgroup of G corresponding to $\pi_1(N_p)$. Then $G'(N_p) = \psi(G(N_p))$ is geometrically finite, hence quasi-Fuchsian.*

PROOF. Similarly to the proof of the preceding lemma, we have only to prove that $\mathbf{H}^3/G'(N_p)$ cannot have a simply-degenerate end. Let $C(N_p)$ be a compact core of $\mathbf{H}^3/G(N_p)$ which is a homeomorphic lift of N_p. Observe that $\mathbf{H}^3/G(N_p)$ is homeomorphic to $T \times \mathbf{R}$ for a closed orientable surface T.

Suppose, seeking a contradiction, that $\mathbf{H}^3/G'(N_p)$ has a simply degenerate end. First, let us consider the case when both of the ends of $\mathbf{H}^3/G'(N_p)$ are simply degenerate. Let V and V' be the components of $V(G)$ that are attached to N_p at the two boundary components of N_p. It is possible that V coincides with V'. Consider the compact submanifold W_p of $C(G)$ which is the union of V, V' and N_p. (In the case when $V = V'$, we mean the compact manifold which is the union of N_p and V.) We define $G(W_p)$ to be a subgroup of G corresponding to the image of $\pi_1(W_p)$ in $G \cong \pi_1(C(G))$, which contains $G(N_p)$, and $G'(W_p)$ to be $\psi(G(W_p))$.

Let $C'(W_p)$ be a compact core of $\mathbf{H}^3/G'(W_p)$. Since W_p is boundary-reducible, the group $G'(W_p)$, which is isomorphic to $G(W_p) \cong \pi_1(W_p)$, is freely decomposable. Therefore $C'(W_p)$ must be also boundary-reducible. Let $p'_{W_p} : \mathbf{H}^3/G'(N_p) \to \mathbf{H}^3/G'(W_p)$ be the covering associated with the inclusion $G'(N_p) \subset G'(W_p)$. By Thurston's covering theorem, either end of $\mathbf{H}^3/G'(N_p)$, which we assumed to be simply degenerate, has a neighbourhood that finitely covers a neighbourhood of an end of $\mathbf{H}^3/G'(W_p)$ by the restriction of p'_{W_p}. It follows that with respect to a product structure $T \times \mathbf{R}$ of $\mathbf{H}^3/G'(N_p)$, there exist $s, t \in \mathbf{R}$ such that both $p'_{W_p}|(T \times (-\infty, s))$ and $p'_{W_p}|(T \times (t, \infty))$ are finite-sheeted covering to its images which are disjoint from $C'(W_p)$.

Because $C'(W_p)$ is boundary-reducible, there is a compressible component F of $\partial C'(W_p)$. Since p'_{W_p} is surjective, either $p'_{W_p}|(T \times (-\infty, s))$ or $p'_{W_p}|(T \times (t, \infty))$, which we can assume to be $p'_{W_p}|(T \times (-\infty, s))$, is contained in the component of $\mathbf{H}^3/G'(W_p) - C'(W_p)$ facing F. Since $p'_{W_p}|(T \times (-\infty, s))$ is a finite-sheeted covering to its image, we can isotope the product structure so that $p'_{W_p}|(T \times \{\sigma\})$ is a covering to its image for $\sigma \in (-\infty, s)$. As $T \times \{\sigma\}$ is incompressible in $\mathbf{H}^3/G'(N_p)$, its image $T' = p'_{W_p}(T \times \{\sigma\})$, whose fundamental group is a finite-indexed extension of $p'_{W_p\#}\pi_1(T \times \{\sigma\})$, is also incompressible. The surface T' can be homotopic to an immersed surface in the compact core $C'(W_p)$.

By the theorem of Freedman-Hass-Scott [**18**], an incompressible immersion which is homotopic to an embedding is homotopic to an embedding whose image is contained in its small regular neighbourhood. Therefore T' is homotopic to an embedded surface $T_0 \subset C'(W_p)$. By classical 3-dimensional topology, $T' \cup T_0$ bounds a 3-manifold X homeomorphic to $T \times I$. This implies that even after we cut W_p along T_0 and discard the part contained in X, we still have a compact core. On the other hand, as we assumed that T' is contained in the component of $\mathbf{H}^3/G'(W_p) - C'(W_p)$ facing F, the compressible boundary component F of $C'(W_p)$ is contained in X. This contradicts the uniqueness of homeomorphism type of compact core.

Next suppose that only one of the ends of $\mathbf{H}^3/G'(N_p)$ is simply degenerate, which means that the other end is geometrically finite. Recall that the boundary components of N_p, which we denote by T_0 and T_1, correspond to the components of $\Omega_{G(N_p)}/G(N_p)$ as $G(N_p)$ is geometrically finite. A quasi-conformal deformation $(\psi_i(G(N_p)), \psi_i|G(N_p))$ determines marked hyperbolic structures m_i^0 on T_0 and m_i^1 on T_1 by this correspondence. Since the compact core $C(N_p)$ is homeomorphic to $T \times I$, we can identify T_1 with $T \times \{1\} \cong T$ naturally by an orientation-preserving homeomorphism. (Here we fix an orientation on $T \times I$.) We regard the hyperbolic structure m_i^0 on T_0 as defining its complex conjugate $\overline{m}_i^0$ on T by the correspondence of T_0 with $T \times \{0\} \cong T$. After taking a subsequence, each of the sequences $\{\overline{m}_i^0\}$ and $\{m_i^1\}$ can be assumed to converge either in the Teichmüller space of T or to a projective lamination in its Thurston compactification.

Suppose that both $\{\overline{m}_i^0\}$ and $\{m_i^1\}$ converge to projective laminations, $[\lambda_0]$ and $[\lambda_1]$ respectively. Then by the same argument as the proof of Lemma 7.3, neither λ_0 nor λ_1 is realized by a pleated surface homotopic to $\Psi^{N_p}|T$, where Ψ^{N_p} is a homotopy equivalence induced from the isomorphism $\psi|G(N_p) : G(N_p) \to G'(N_p)$ and T is identified with $T \times \{1\}$. It is known, as was noted in Thurston [**45**] (see Lemma 4.1 in Ohshika [**34**]), that each measured lamination must be maximal and connected as we assumed that G', hence also $G'(N_p)$, has no parabolic elements. Furthermore, as was shown in Ohshika [**37**], it is impossible that λ_0 and λ_1 have the same support. It follows by the uniqueness of support of ending lamination (Thurston [**45**], see also Ohshika [**34**],) that $\mathbf{H}^3/G'(N_p)$ has two simply-degenerate ends, whose ending laminations are represented by $\Psi^{N_p}(\lambda_0)$ and $\Psi^{N_p}(\lambda_1)$ respectively. This contradicts our assumption here that $\mathbf{H}^3/G(N_p)$ has only one simply-degenerate end.

Thus we can assume that at least one of $\{\overline{m}_i^0\}$ and $\{m_i^1\}$ converges inside the Teichmüller space after taking a subsequence. If both of them converge inside the Teichmüller space, then the limit $G'(N_0)$ is quasi-Fuchsian, contradicting our assumption. Hence we can assume that exactly one of them, say $\{m_i^1\}$, converges inside the Teichmüller space after taking a subsequence. We denote a convergent subsequence by the same symbol $\{m_i^1\}$.

Since $C(N_p)$ is mapped homeomorphically to N_p by the covering projection, the boundary components T_0 and T_1 of $C(N_p)$ are mapped to those of N_p, which we denote by $\overline{T}_0$ and $\overline{T}_1$. Let V be a component of $V(G)$ which contains $\overline{T}_0$ as a boundary component. Let W_p be the compact submanifold of $C(G)$ which is the union of N_p and V. As before, let $G(W_p)$ be a subgroup of G corresponding to the image of $\pi_1(W_p)$ in G, which contains $G(N_p)$ as a subgroup. The submanifold W_p is lifted homeomorphically to a compact core $C(W_p)$ of $\mathbf{H}^3/G(W_p)$. The boundary component $\overline{T}_1$ of W_p is lifted to a boundary component $\tilde{T}_1$ of $C(W_p)$.

The subgroup $\psi_i(G(W_p))$, which we denote by $G_i(W_p)$, of G_i is a quasi-conformal deformation of $G(W_p)$. We denote $\psi_i|G(W_p)$ by $\psi_i^{W_p}$ and $\psi|G(W_p)$ by ψ^{W_p}. A homotopy equivalence $\Psi_i^{W_p}$ can be assumed to map $C(W_p)$ homeomorphically to a compact core $C_i(W_p)$ of $\mathbf{H}^3/G_i(W_p)$. By our definition of W_p, the Riemann surface $\Omega_{G_i(W_p)}/G_i(W_p)$ has a component $\hat{T}_1(i)$ corresponding to the boundary at infinity in the Kleinian manifold for a geometrically finite end facing a boundary component $T_1(i)$ of $C_i(W_p)$ which is homotopic to $\Psi_i^{W_p}(\tilde{T}_1)$. The marked hyperbolic structure on T_1, which is induced from the conformal structure on $\hat{T}_1(i)$ by the composition of $\Psi_i^{W_p}$ and the covering projection, coincides with m_i^1 by our construction of W_p. As we assumed that $\{m_i^1\}$ converges inside the Teichmüller space, for the algebraic limit $(G'(W_p), \psi^{W_p})$ of $\{G_i(W_p), \psi_i^{W_p})\}$, which is contained in G' and contains $G'(N_p)$ as a subgroup, the hyperbolic 3-manifold $\mathbf{H}^3/G'(W_p)$ has a geometrically finite end facing a boundary component T_1' of a compact core $C'(W_p)$ homotopic to $\Psi^{W_p}(\tilde{T}_1)$ by the same argument as Lemma 7.2.

On the other hand, by assumption, $\mathbf{H}^3/G'(N_p)$ has a simply-degenerate end with a neighbourhood parametrized by $T \times (0, \infty)$. Let $p'_{W_p} : \mathbf{H}^3/G'(N_p) \to \mathbf{H}^3/G'(W_p)$ be the covering associated with the inclusion as before. By Thurston's covering theorem, we can take such a neighbourhood so that $p'_{W_p}|T \times (0, \infty)$ is a finite-sheeted covering to its image. Hence for a sufficiently large $t \in (0, \infty)$, the image $p'_{W_p}(T \times (t, \infty))$ is disjoint from the compact core $C'(W_p)$ and contained in a component of its complement containing a geometrically infinite end. Since both $p'_{W_p}(T \times \{t\})$ and T_1' are homotopic to $\Psi^{W_p}(\tilde{T}_1)$, they are homotopic. By the argument as before, $p'_{W_p}(T \times \{t\})$ can be homotoped to an embedding, which we denote by T_0', contained in the neighbourhood of its image, and T_0' and T_1' cobound a submanifold homeomorphic to $T \times I$, which contains $C'(W_p)$ inside. This is a contradiction.

Thus we have proved that $\mathbf{H}^3/G'(N_p)$ cannot have a simply-degenerate end, hence $G'(N_p)$ must be geometrically finite. □

The following Lemma 8.8 constitutes the main part of the proof of Lemma 8.2. It will be used to show that Ψ can be homotoped so that the restriction of Ψ to the union of the components of $N(G)$ which are not trivial I-bundles over closed surfaces, and its composition with the projection to the geometric limit $\mathbf{H}^3/G_\infty$ are embeddings.

To prove Lemma 8.8, we shall construct a submanifold homotopic to $\Psi_i|N_0$ in the convex core of $\mathbf{H}^3/G_i$ so that its diameter and the distance from the basepoint are bounded as $i \to \infty$. For that, using Maskit's combination theorem, we shall decompose the convex core of $\mathbf{H}^3/G_i$ into parts corresponding to components of $N(G)$ and $V(G)$, and then compress the boundary of the part corresponding to N_0 within a universally bounded distance from the basepoint.

Now let us state Lemma 8.8.

LEMMA 8.8. *Let N_0 be a component of $N(G)$ which is not a product I-bundle over a closed surface. Let $p'_{N_0} : \mathbf{H}^3/G'(N_0) \to \mathbf{H}^3/G'$ be the covering map associated with the inclusion $G'(N_0) \subset G'$. Then there exists a compact core $C'(N_0)$ of $\mathbf{H}^3/G'(N_0)$ such that $p'_{N_0}|C'(N_0)$ is an embedding. Similarly, if N_0 and $N_{0'}$ are two distinct components of $N(G)$ neither of which is a product I-bundle over a closed surface, then we can take compact cores $C'(N_0)$ of $\mathbf{H}^3/G'(N_0)$ and $C'(N_{0'})$ of $\mathbf{H}^3/G'(N_{0'})$ as above such that $p'_{N_0}(C'(N_0))$ and $p'_{N_{0'}}(C'(N_{0'}))$ are mutually disjoint. Furthermore, let G_∞ be the geometric limit of $\{G_i\}$ and let $p_\infty : \mathbf{H}^3/G' \to \mathbf{H}^3/G_\infty$ be the covering associated with the inclusion. Then the two assertions above remain true after replacing p'_{N_0} with $p_\infty \circ p'_{N_0}$, and $p'_{N_{0'}}$ with $p_\infty \circ p'_{N_{0'}}$.*

Before starting the proof, we fix a system of the symbols according to the following convention:

CONVENTION 2. Let F be an incompressible surface in N_0. A subgroup of G corresponding to $\pi_1(F)$ is denoted by $G(F)$. For G' and G_i, the corresponding groups are denoted by $G'(F)$ and $G_i(F)$ respectively. A homeomorphic lift of an incompressible surface F in $\mathbf{H}^3/G(N_0)$ is denoted by $\hat{F}$, and one in $\mathbf{H}/G(F)$ is denoted by $\tilde{F}$. When we need to consider a component V_0 of $V(G)$ containing F and a subgroup $G(V_0)$ associated with the image of $\pi_1(V_0)$, we denote a lift of a surface F to $\mathbf{H}^3/G(V_0)$ by $\dot{F}$. The same rule applies for G_i and G'. The lift of a homotopy equivalence, e.g., $\Psi : \mathbf{H}^3/G \to \mathbf{H}^3/G'$ to a map between coverings, e.g., from $\mathbf{H}^3/G(F)$ to $\mathbf{H}^3/G'(F)$ is denoted by Ψ^F.

Also, we shall use the following notion of visual function and its property:

DEFINITION 8.9. Let Ω be a domain in S^2_∞. For each point $x \in \mathbf{H}^3$, we define the *visual measure* of Ω at x to be the two-dimensional Lebesgue measure of the image of Ω on the unit tangent obtained by projecting Ω along geodesic rays issued at x. The function whose value at x is equal to the visual measure of Ω at x is called the *visual measure function* of Ω.

LEMMA 8.10. *For any domain on S^2_∞, its visual measure function is a smooth harmonic function.*

Refer to §7 of Canary [**7**] for details of this fact.

We now start the proof of the lemma.

PROOF OF LEMMA 8.8. Suppose that Σ is an internal boundary component of N_0. Then there is a simply-connected component $\Omega(\Sigma)$ of the region of discontinuity $\Omega_{G(N_0)}$, which is invariant by $G(\Sigma)$ as Σ is incompressible. Similarly, there is a component $\Omega_i(\Sigma)$ of the region of discontinuity $\Omega_{G_i(N_0)}$, which is invariant by $G_i(\Sigma)$.

Let $\tilde{v}_i : \mathbf{H}^3 \to \mathbf{R}$ be the visual measure function of $\Omega_i(\Sigma)$ as defined in Definition 8.9, which is harmonic. Since $\Omega_i(\Sigma)$ is $G_i(\Sigma)$-invariant, the function $\tilde{v}_i$ is also $G_i(\Sigma)$-invariant, hence induces a harmonic function v_i defined on $\mathbf{H}^3/G_i(\Sigma)$.

For a given constant ϵ, let $D_i(\epsilon)$ be the subset of $\mathbf{H}^3$ consisting of the points where the value of $\tilde{v}_i$ is at least $2\pi - \epsilon$. Then we get the following lemma.

LEMMA 8.11. *Suppose that N_0 is not an I-bundle over a closed surface. Then for each i, there exists a positive constant ϵ depending on i such that $D_i(\epsilon)$ meets none of its translates by elements of $G_i(N_0) - G_i(\Sigma)$.*

PROOF. Let γ be an element in $G_i(N_0) - G_i(\Sigma)$. Then $\gamma\Omega_i(\Sigma) \cap \Omega_i(\Sigma) = \emptyset$ because the stabilizer of $\Omega_i(\Sigma)$ is equal to $G_i(\Sigma)$. Let $\Omega(\gamma)$ be the union of the translates of $\gamma\Omega_i(\Sigma)$ by the elements of $G_i(\Sigma)$. Let $D_\gamma(\epsilon)$ be the subset of $\mathbf{H}^3$ consisting of the points where the visual measure of $\Omega(\gamma)$ is at least $2\pi - \epsilon$. Obviously $D_\gamma(\epsilon)$ contains $\gamma D_i(\epsilon)$.

Now, since we assumed that N_0 is not an I-bundle over a closed surface, there exist more than two cosets in $G_i(\Sigma) \setminus G_i(N_0)$. (Recall that a compact, orientable, irreducible 3-manifold has fundamental group which is an extension of a closed surface group of index 2 if and only if it is homeomorphic to a twisted I-bundle of a closed non-orientable surface.) Take a $\gamma' \in G_i(N_0) - G_i(\Sigma)$ which is contained in a coset different from that of γ. Let $\Omega(\gamma')$ be the union of the translates of $\gamma'\Omega_i(\Sigma)$ by the elements of $G_i(\Sigma)$. Then we have $\Omega(\gamma) \cap \Omega(\gamma') = \emptyset$ because γ and γ' belong to different cosets of $G_i(\Sigma) \setminus G_i(N_0)$ and the stabilizer of $\Omega_i(\Sigma)$ is equal to $G_i(\Sigma)$.

Let $\tilde{v}_\gamma$ be the visual measure function of $\Omega(\gamma)$, and let $\tilde{v}_{\gamma'}$ be that of $\Omega(\gamma')$. Since both $\Omega(\gamma)$ and $\Omega(\gamma')$ are $G_i(\Sigma)$-invariant, the functions $\tilde{v}_\gamma$ and $\tilde{v}_{\gamma'}$ induce harmonic functions v_γ and $v_{\gamma'}$ on $\mathbf{H}^3/G_i(\Sigma)$ respectively.

We proved that the Kleinian group $G_i(\Sigma)$ is geometrically finite in Lemma 8.6. Let C_i be the convex core of $\mathbf{H}^3/G_i(\Sigma)$. Then $\mathbf{H}^3/G_i(\Sigma)$ has a geometrically finite end facing a boundary component S_i of C_i whose boundary at infinity is $\Omega_i(\Sigma)/G_i(\Sigma)$. The function v_i takes value greater than 2π in the component of $\mathbf{H}^3/G_i(\Sigma) - C_i$ facing S_i, and as a point tends to the end corresponding to $\Omega_i(\Sigma)/G_i(\Sigma)$, the value of v_i approaches to 4π. On the other hand, the value of v_i goes to 0 as a point tends to the other end.

Let $T(\epsilon_0)$ be the subset of $\mathbf{H}^3/G_i(\Sigma)$ consisting of the points x such that $2\pi - \epsilon_0 \leq v_i(x) \leq 2\pi + \epsilon_0$ for a fixed small $\epsilon_0 > 0$. If we take a sufficiently small ϵ_0, then the set $T(\epsilon_0)$ is compact by the observation above and the assumption that G_i has no parabolic elements. Since $v_{\gamma'}$ is a positive continuous function, it attains a minimum value $\epsilon_1 > 0$ in $T(\epsilon_0)$. Let ϵ_2 be $\min\{\epsilon_1/3, \epsilon_0\}$. We shall prove by contradiction that $D_i(\epsilon_2)$ and $D_\gamma(\epsilon_2)$ do not intersect each other.

Suppose that $D_\gamma(\epsilon_2)$ and $D_i(\epsilon_2)$ intersect, and let x be a point in $D_\gamma(\epsilon_2) \cap D_i(\epsilon_2)$. Let $\tilde{T}(\epsilon_0)$ be the preimage of $T(\epsilon_0)$ in $\mathbf{H}^3$ with respect to the universal covering projection from $\mathbf{H}^3$ to $\mathbf{H}^3/G_i(\Sigma)$. Since $\tilde{v}_i(x) \geq 2\pi - \epsilon_2 \geq 2\pi - \epsilon_0$, we have either $x \in \tilde{T}(\epsilon_0) \cap D_\gamma(\epsilon_2) \cap D_i(\epsilon_2)$ or $\tilde{v}_i(x) > 2\pi + \epsilon_0 \geq 2\pi + \epsilon_2$. In the former case, we have $\tilde{v}_{\gamma'}(x) \geq \epsilon_1, \tilde{v}_\gamma(x) \geq 2\pi - \epsilon_2 \geq 2\pi - \epsilon_1/3$, and $\tilde{v}_i(x) \geq 2\pi - \epsilon_0 \geq 2\pi - \epsilon_1/3$. This contradicts the facts that the total visual measure of S^2_∞ is 4π and that $\Omega(\gamma), \Omega(\gamma')$ and $\Omega_i(\Sigma)$ are mutually disjoint. In the latter case, we have $\tilde{v}_i(x) > 2\pi + \epsilon_2$ and $\tilde{v}_\gamma(x) \geq 2\pi - \epsilon_2$, which also contradicts the fact that the total visual measure of S^2_∞ is 4π. Thus we have proved that $D_\gamma(\epsilon_2) \cap D_i(\epsilon_2) = \emptyset$.

Note that the constant ϵ_2 above depends on the choice of the coset containing γ' (and on i) but is independent of γ. Consequently, for every $\gamma \in G$, except those in the coset containing γ', we have $D_\gamma(\epsilon_2) \cap D_i(\epsilon_2) = \emptyset$. For elements contained in the coset of γ', we can obtain another positive constant ϵ_3 such that $D_{\gamma'}(\epsilon_3) \cap D_i(\epsilon_3) = \emptyset$, by interchanging the roles of γ and γ' in the argument above. Hence by setting $\epsilon = \min\{\epsilon_2, \epsilon_3\}$, we complete the proof. □

REMARK 6. Even when N_0 is a twisted I-bundle over a closed non-orientable surface, by a similar but simpler argument, we can prove that the interior of $D_i(0)$ does not intersect its translates by the elements of $G_i(N_0) - G_i(\Sigma)$.

Also the same argument as Lemma 8.11 works for a component V_0 of $V(G)$. Let $G(V_0)$ be a subgroup of G corresponding the image of $\pi_1(V_0)$ by the homomorphism induced from the inclusion of $V_0 \subset \mathbf{H}^3/G$, and let $G_i(V_0)$ be $\psi_i(G(V_0))$. Let Σ be an incompressible boundary component of V_0. Let $G(\Sigma)$ be a subgroup of $G(V_0)$ corresponding to the image of $\pi_1(\Sigma) \subset \pi_1(V_0)$ and let $G_i(\Sigma)$ be $\psi_i(G(\Sigma))$. Then the region of discontinuity $\Omega_{G_i(V_0)}$ contains a component $\Omega'_i(\Sigma)$ whose stabilizer is equal to $G_i(\Sigma)$.

Similarly to before, we can consider the visual measure function $\tilde{v'}_i : \mathbf{H}^3 \to \mathbf{R}$ of $\Omega'_i(\Sigma)$, and the subset $D'_i(\epsilon)$ consisting of the points where the value of $\tilde{v'}_i$ is at least $2\pi - \epsilon$ for an $\epsilon > 0$. Then by the same argument as the proof of Lemma 8.11, we have the following.

LEMMA 8.12. *There exists a constant $\epsilon > 0$ depending on i such that $D'_i(\epsilon)$ intersects none of its translates by elements of $G_i(V_0) - G_i(\Sigma)$.*

Now suppose that V_0 and N_0 have a common boundary component Σ. Let W_0 be the compact 3-sub-manifold which is obtained by pasting V_0 and

N_0 along Σ. Let $G(W_0)$ be a subgroup of G corresponding to the image of $\pi_1(W_0)$. Then the subgroup $G_i(W_0)$ of G_i which we define to be $\psi_i(G(W_0))$ can be obtained by Maskit's combination from $G_i(V_0)$ and $G_i(N_0)$. (As abstract groups, $G_i(W_0)$ is isomorphic to the amalgamated free product of $G_i(N_0)$ and $G_i(V_0)$ on $G_i(\Sigma)$.)

As is shown in Theorem 8.2 of Morgan [**33**], the convex core of $G_i(W_0)$ can be constructed explicitly from those of $G_i(N_0)$ and $G_i(V_0)$ as follows by using the results above. First we can take $G(N_0)$ and $G(V_0)$ so that $G(N_0) \cap G(V_0)$ contains $G(\Sigma)$. Let $p^i_{(\Sigma,N_0)} : \mathbf{H}^3/G_i(\Sigma) \to \mathbf{H}^3/G_i(N_0)$ and $p^i_{(\Sigma,V_0)} : \mathbf{H}^3/G_i(\Sigma) \to \mathbf{H}^3/G_i(V_0)$ be the coverings associated with the inclusions. Then by Lemma 8.11 and the remarks above, we can see that $p^i_{(\Sigma,N_0)}|(\mathrm{Int}D_i(0)/G_i(\Sigma))$ and $p^i_{(\Sigma,V_0)}|(D'_i(0)/G_i(\Sigma))$ are injective.

On the other hand, $\Omega_{G_i(\Sigma)}$ consists of two components $\Omega^0_{G_i(\Sigma)}$ and $\Omega^1_{G_i(\Sigma)}$ such that $\Omega^0_{G_i(\Sigma)}$ is conformal to $\Omega_i(\Sigma)$ and $\Omega^1_{G_i(\Sigma)}$ is conformal to $\Omega'{}_i(\Sigma)$ equivariantly with respect to the actions of $G_i(\Sigma)$. Since the sum of the visual measures of $\Omega^0_{G_i(\Sigma)}$ and $\Omega^1_{G_i(\Sigma)}$ is constantly 4π, we can see that the sets $\{x|v_i(x) = 2\pi - \epsilon\}$ and $\{x|v'_i(x) = 2\pi + \epsilon\}$ correspond one to one equivariantly with respect to the action of $G_i(\Sigma)$. The constant ϵ is allowed to take both positive and negative values provided that its absolute value is small enough in the case when N_0 is not an I-bundle over a closed surface. When N_0 is a twisted I-bundle over a non-orientable closed surface, ϵ should be negative.

In particular, $\mathrm{Fr}p^i_{(\Sigma,N_0)}(D_i(0)/G_i(\Sigma))$ and $\mathrm{Fr}p^i_{(\Sigma,V_0)}(D'_i(0)/G_i(\Sigma))$ are isometric unless N_0 is an I-bundle over a closed surface. (Note that $\mathrm{Fr}p^i_{(\Sigma,N_0)}(D_i(0)/G_i(\Sigma))$ may have singularities however. The precise meaning of isometry here is that between neighbourhoods as will be explained below.) The convex core of $\mathbf{H}^3/G_i(W_0)$ is obtained by pasting $C_{G_i(N_0)} - p^i_{(\Sigma,N_0)}\{\mathrm{Int}(D_i(0)/G_i(\Sigma))\}$ and $C_{G_i(V_0)} - p^i_{(\Sigma,V_0)}\{\mathrm{Int}(D'_i(0)/G_i(\Sigma))\}$ along their frontiers. More precisely, we identify

$$\mathrm{Int}p^i_{(\Sigma,N_0)}\{(D_i(\epsilon)-D_i(-\epsilon))/G_i(\Sigma)\} \text{ in } C_{G_i(N_0)}-p^i_{(\Sigma,N_0)}\{\mathrm{Int}(D_i(-\epsilon)/G_i(\Sigma))\}$$

with

$$\mathrm{Int}p^i_{(\Sigma,V_0)}\{(D'_i(\epsilon)-D'_i(-\epsilon))/G_i(\Sigma)\} \text{ in } C_{G_i(V_0)}-p^i_{(\Sigma,V_0)}\{\mathrm{Int}(D'_i(-\epsilon)/G_i(\Sigma))\}$$

by an isometry for some small positive ϵ for which both of the above are embeddings without singularity on the frontiers. If N_0 is a twisted I-bundle over a non-orientable surface, then we can identify $\mathrm{Int}p^i_{(\Sigma,N_0)}\{((D_i(-\epsilon/2) - D_i(-\epsilon))/G_i(\Sigma)\}$ with $\mathrm{Int}p^i_{(\Sigma,V_0)}\{(D'_i(\epsilon)-D'_i(\epsilon/2))/G_i(\Sigma)\}$ by an isometry for a positive ϵ for which both of the above are embeddings without singularity on the frontiers. We can prove such a positive number ϵ exists by Sard's theorem since the visual measure function is smooth.

We can construct the convex core of $\mathbf{H}^3/G_i$ from the convex cores of $\mathbf{H}^3/G_i(N_k)$ and $\mathbf{H}^3/G_i(V_l)$ for the components N_k of $N(G)$ and V_l of $V(G)$

in a similar way as follows. For each component N_k of $N(G)$ that is not a product I-bundle, let $G(N_k)$ be a subgroup of G corresponding to the image of $\pi_1(N_k)$ and let $G_i(N_k)$ be $\psi_i(G(N_k))$. In the same fashion, for each component V_l of $V(G)$, let $G(V_l)$ be a subgroup of G corresponding to the image of $\pi_1(V_l)$ and let $G_i(V_l)$ be $\psi_i(G(V_l))$. The Kleinian group G_i is obtained by Maskit's combination from the $G_i(N_k)$'s and the $G_i(V_l)$'s. Then, similarly to the construction of the convex core of $\mathbf{H}^3/G_i(W_0)$ above, the convex core of $\mathbf{H}^3/G_i$ is constructed by pasting subsets of the convex cores of $\mathbf{H}^3/G_i(N_k)$ and $\mathbf{H}^3/G_i(V_l)$ as follows.

Let Σ_j be a boundary component of N_k which is also contained in the boundary of $V(G)$. We define $G_i(\Sigma_j)$ by exactly the same way as before. Let $\Omega_i(\Sigma_j)$ be the component of the region of discontinuity $\Omega_{G_i(N_k)}$ whose stabilizer is $G_i(\Sigma_j) \subset G_i(N_k)$. We can define as before the subset $D_i^j(\epsilon)$ of $\mathbf{H}^3$ consisting of the points where the visual measure of $\Omega_i(\Sigma_j)$ is at least $2\pi - \epsilon$. As was shown in Lemma 8.11 and the remark following it, the interior of $D_i^j(0)$ intersects none of its translates by the elements of $G_i(N_k) - G_i(\Sigma_j)$. If $j \neq j'$ (note that this happens only when N_k is not an I-bundle over a closed surface), then $D_i^j(0)$ intersects no translates of $D_i^{j'}(0)$ by elements of $G_i(N_k)$ because (1) $\Omega_i(\Sigma_k)$ intersects no translates of $\Omega_i(\Sigma_{j'})$; (2) the region of discontinuity $\Omega_{G_i(N_k)}$ has more than two components; and (3) the sum of the visual measures of disjoint components cannot exceed 4π.

Similarly, we define for each incompressible boundary component Σ'_j of V_l, the subset $D'^j_i(\epsilon)$ of $\mathbf{H}^3$ consisting of the points where the visual measure of $\Omega_i(\Sigma'_j)$, which is a component of $\Omega_{G_i(V_l)}$ invariant under $G_i(\Sigma'_j)$, is at least $2\pi - \epsilon$. Then $D'^j_i(0)$ is invariant under $G_i(\Sigma'_j)$ and intersects none of its translates by the elements in $G_i(V_l) - G_i(\Sigma'_j)$ as shown in Lemma 8.12. Also in this case, if $j \neq j'$, then $D'^j_i(0)$ intersects no translates of $D'^{j'}_i(0)$ by elements of $G_i(V_l)$.

Let $p^i_{(\Sigma_j, N_k)} : \mathbf{H}^3/G_i^{\Sigma_j} \to \mathbf{H}^3/G_i(N_k)$ and $p^i_{(\Sigma'_j, V_l)} : \mathbf{H}^3/G_i(\Sigma'_j) \to \mathbf{H}^3/G_i(V_l)$ be the covering projections associated with the inclusions. Then $p^i_{(\Sigma_j,N_k)}|\mathrm{Int}(D_i^j(0)/G_i(\Sigma_j))$ and $p^i_{(\Sigma'_j,V_l)}|(D'^j_i(0)/G_i(\Sigma'_j))$ are injective. We can construct the convex core of $\mathbf{H}^3/G_i$ by pasting

$$\{C_{G_i(N_k)} - \cup_j p^i_{(\Sigma_k,N_k)}\mathrm{Int}((D_i^j(0)/G_i(\Sigma_j))\}$$

and

$$\{C_{G_i(V_l)} - \cup_j p^i_{(\Sigma'_j,V_l)}(\mathrm{Int}(D'^j_i(0)/G_i(\Sigma'_j))\}$$

along their frontiers in such a way that components corresponding to the same surface in $C(G)$ are matched. (For each Σ_j there is a corresponding Σ'_j. As before, more precisely, the pasting here means that we identify either

$$p^i_{(\Sigma_j,N_k)}((D_i^j(\epsilon) - D_i^j(-\epsilon))/G_i(\Sigma_j))$$

with

$$p^i_{(\Sigma'_j,V_l)}((D'^j_i(\epsilon) - D'^j_i(-\epsilon))/G_i(\Sigma'_j)),$$

or

$$p^i_{(\Sigma_j,N_k)}(D^j_i(-\epsilon/2) - D^j_i(-\epsilon))/G_i(\Sigma_j)$$

with

$$p^i_{(\Sigma'_j,V_l)}(D'^j_i(\epsilon) - D'^j_i(\epsilon/2))/G_i(\Sigma'_j)$$

in the case when N_k is a twisted I- bundle.)

By Lemma 8.11, unless N_k is an I-bundle over a closed surface, $p^i_{(\Sigma_k,N_k)}|(D^j_i(\epsilon)/G_i(\Sigma_j))$ is an embedding for a sufficiently small positive constant ϵ. Then we can take a positive ϵ_i satisfying the conclusion of Lemma 8.11 which goes to 0 as $i \to \infty$ so that $2\pi - \epsilon_i$ is a regular value of the visual measure function and that $p^i_{(\Sigma_j,N_k)}|(D^j_i(\epsilon_i)/G_i(\Sigma_j))$ is an embedding for every k and j. Let $p_i^{N_k} : \mathbf{H}^3/G_i(N_k) \to \mathbf{H}^3/G_i$ be the covering associated to the inclusion. Then for $Q_i^k = C_{G_i(N_k)} - \cup_j p^i_{(\Sigma_j,N_k)}(\mathrm{Int} D^j_i(\epsilon_i)/G_i(\Sigma_j))$, the restriction of the covering projection $p_i^{N_k}|Q_i^k$ is an embedding into the convex core C_{G_i}, and $p_i^{N_k}(Q_i^k)$ and $p_i^{N_{k'}}(Q_i^{k'})$ are mutually disjoint if $k \neq k'$, as we can see by the way of constructing C_{G_i} using Maskit's combination as explained above.

We need to perform surgery for the compact submanifold Q_i^k to make the boundary incompressible keeping its diameter bounded above as $i \to \infty$. To do that, we need the following lemma. Fix a point $z \in \mathbf{H}^3$. Let x_i be the basepoint in $\mathbf{H}^3/G_i$ that is the image of z by the universal covering projection.

LEMMA 8.13. *Suppose that a component N_k of $N(G)$ is not an I-bundle over a closed surface. Then there exists a constant K independent of i as follows. Let Σ be an internal boundary component of N_k, i.e., one which is not contained in $\partial C(G)$, (see Definition 8.5). Let F_i be the boundary component of $p_i^{N_k}(Q_i^k)$ corresponding to Σ. Then we can compress F_i in $\mathbf{H}^3/G_i$ to an embedded incompressible surface $\overline{F}_i$ homotopic to $\Psi_i(\Sigma)$, which is contained in the K-neighbourhood of x_i. Furthermore we can take such an embedded incompressible surfaces so that the two surfaces corresponding to distinct internal boundary components Σ and Σ^0 of N_k are mutually disjoint.*

PROOF. By the way of constructing Q_i^k, we can see that F_i is a homeomorphic image of a closed embedded surface $\tilde{F}_i$ in $\mathbf{H}^3/G_i(\Sigma)$ by the projection $p_i^{N_k} \circ p^i_{(\Sigma,N_k)}$. A compressing disc for F_i can be lifted to that for $\tilde{F}_i$. Recall that $\tilde{F}_i$ is the boundary of $D_i(\epsilon_i)/G_i(\Sigma)$ for a small positive number ϵ_i.

By choosing a sufficiently small positive ϵ_i for each i, we can assume that $\tilde{F}_i$ is contained in the δ_i-neighbourhood of the frontier of $D_i(0)/G_i(\Sigma)$, where $\delta_i \to 0$ as $i \to \infty$. Then $\tilde{F}_i$ is contained in the δ_i-neighbourhood of the convex core of $\mathbf{H}^3/G_i(\Sigma)$ because the frontier of $D_i(0)/G_i(\Sigma)$ is contained in the convex core of $\mathbf{H}^3/G_i(\Sigma)$. Since the δ_i-neighbourhood of the convex core, which is compact, is also a compact core, any compressing disc of $\tilde{F}_i$ that is a lift of compressing disc of F_i can be homotoped fixing the boundary to a compressing disc contained in the δ_i-neighbourhood of the convex core.

Let Δ_i be such a compressing disc for $\tilde{F}_i$ in the δ_i-neighbourhood of the convex core. Although the disc Δ_i may not be mapped to an embedded disc by the covering projection $p_i^{N_k} \circ p^i_{(\Sigma,N_k)}$, as it is homotopic to the original compressing disc which is an embedding, we can see that there exists a homotopic embedded disc $\overline{\Delta}_i$ arbitrarily near to $p_i^{N_k} \circ p^i_{(\Sigma,N_k)}(\Delta_i)$ as follows. The analytic loop theorem of Meeks-Yau [**31**] claims that in a complete Riemannian 3-manifold, any disc whose boundary does not contain singularities can be homotoped to a an embedded surface of least area relative to the boundary. We consider a Riemannian metric on $\mathbf{H}^3/G_i$ with a heavy weight near the image of Δ_i. Then any least-area surface homotopic to Δ_i must be contained in a neighbourhood of the image of Δ_i. Therefore, we have an embedded surface homotopic to Δ_i contained in an arbitrarily small neighbourhood of Δ_i. Hence by letting a lift of $\overline{\Delta}_i$, whose boundary is on $\tilde{F}_i$, be a new Δ_i, we can assume that Δ_i is mapped to an embedded disc without changing the condition that it is contained in the δ_i-neighbourhood of the convex core.

Suppose that $\overline{\Delta}_i$ intersects F_i outside $\partial\overline{\Delta}_i$. Make the intersection transverse and suppose that we still have extra intersection. Then, the innermost sub-disc bounded by F_i on $\overline{\Delta}_i$ is either a compressing disc for F_i or an inessential disc which we can remove by an isotopy inside the image of the δ_i-neighbourhood of the convex core. Anyway, as a sub-disc of $\overline{\Delta}_i$ or a disc obtained by isotoping $\overline{\Delta}_i$, we get a compressing disc for F_i. We change $\overline{\Delta}_i$ to such a compressing disc. We also change Δ_i to a lift of $\overline{\Delta}_i$. We compress F_i along such a compressing disc $\overline{\Delta}_i$. Since F_i is a closed surface, by repeating surgery along such compressing discs, we get an embedded incompressible surface $\overline{F}_i$ homotopic to $\Psi_i(\Sigma)$ whose lift to $\mathbf{H}^3/G_i(\Sigma)$ is contained in the δ_i-neighbourhood of the convex core of $\mathbf{H}^3/G_i(\Sigma)$.

By Lemma 8.6, $\{(G_i(\Sigma), \psi_i|G(\Sigma))\}$ converges to a quasi-Fuchsian group $(G'(\Sigma), \psi|G(\Sigma))$. Then there is a K_i-bi-Lipschitz diffeomorphism from $\mathbf{H}^3/G_i(\Sigma)$ to $\mathbf{H}^3/G'(\Sigma)$ preserving the basepoints with $K_i \to 1$. This implies that the convex cores of $\mathbf{H}^3/G_i(\Sigma)$ converge geometrically to that of $\mathbf{H}^3/G'(\Sigma)$ as was seen in the proof of Lemma 6.5. It follows that the maximal distance between the basepoint, which is the image of $z \in \mathbf{H}^3$, and the points in the δ_i-neighbourhood of the convex core of $\mathbf{H}^3/G_i(\Sigma)$ must be bounded as $i \to \infty$ because the convex core of the strong limit has bounded diameter as $G'(\Sigma)$ is a geometrically finite group without parabolic elements. Thus we have proved that there exists a constant K such that F_i is compressed to an incompressible surface $\overline{F}_i$ homotopic to $\Psi_i(\Sigma)$ which is contained in the K-neighbourhood of the basepoint.

Perform this surgery first for an internal boundary component Σ without considering the other boundary components. If compressing discs used in the operation intersect another F_i^0 corresponding to another internal boundary component Σ^0 inessentially, we can isotope F_i within a bounded distance to make the disc disjoint from it. If there is a compressing disc intersecting F_i essentially, i.e., by a compressing disc, we use the latter disc to compress F_i afterward so that the obtained incompressible surfaces would be mutually disjoint.

Taking this into consideration, we continue to perform surgery for other boundary components by compressing discs which are disjoint from the surfaces on which the surgery has already been performed. The incompressible surfaces that we have obtained by the surgery are contained in the K-neighbourhood of x_i for a constant K independent of i, since the compressing discs were taken inside the K-neighbourhood of x_i as was shown above. □

By using Lemma 8.13, we can show that there is an embedded incompressible surface in $\mathbf{H}^3/G'$ homotopic to $\Psi(\Sigma)$ as follows if Σ is an internal boundary component of a component of $N(G)$ which is not an I-bundle over a closed surface.

LEMMA 8.14. *Suppose that N_k is a component of $N(G)$ which is not an I-bundle over a closed surface. Let Σ be an internal boundary component of N_k. Then there exists an embedded incompressible surface F' in $\mathbf{H}^3/G'$ which is homotopic to $\Psi(\Sigma)$ as images, i.e., forgetting the markings. Moreover we can choose such an embedded incompressible surface so that two surfaces F' and $F^{0\prime}$ corresponding to distinct internal boundary components Σ and Σ^0 are mutually disjoint.*

PROOF. Recall that G_∞ is the geometric limit of $\{G_i\}$. Let $x_\infty \in \mathbf{H}^3/G_\infty$ be the basepoint of $\mathbf{H}^3/G_\infty$ which is the image of the point $z \in \mathbf{H}^3$ by the universal covering map. Then for each i, there exists a $(1+\epsilon_i, r_i)$-approximate isometry $\rho_i : B_{r_i}(\mathbf{H}^3/G_i, x_i) \to B_{r_i}(\mathbf{H}^3/G_\infty, x_\infty)$, where $\epsilon_i \to 0$ and $r_i \to \infty$ as $i \to \infty$.

Take a sufficiently large integer i_0 so that $r_{i_0} \geq 2K$ for the constant K given in the last lemma. Then for $i > i_0$, the image $\rho_i(\overline{F}_i)$ is an embedded surface in $\mathbf{H}^3/G_\infty$. Let $\tilde{\overline{F}}_i$ be the homeomorphic lift of $\overline{F}_i$ to $\mathbf{H}^3/G_i(\Sigma)$. Let $p_i^\Sigma : \mathbf{H}^3/G_i(\Sigma) \to \mathbf{H}^3/G_i$ be the covering associated with the inclusion. Then the sequence of coverings $p_i^\Sigma : \mathbf{H}^3/G_i(\Sigma) \to \mathbf{H}^3/G_i$ converges geometrically to the covering $p_\infty^\Sigma : \mathbf{H}^3/G'(\Sigma) \to \mathbf{H}^3/G_\infty$ associated to the inclusion $G'(\Sigma) \subset G_\infty$. This means that there is a $(1+\tilde{\epsilon}_i, s_i)$-approximate isometry $\tilde{\rho}_i : B_{s_i}(\mathbf{H}^3/G_i(\Sigma), \tilde{x}_i) \to B_{s_i}(\mathbf{H}^3/G'(\Sigma), \tilde{x}_\infty)$ with $\tilde{\epsilon}_i \to 0$ and $s_i \to \infty$ as $i \to \infty$, such that $\rho_i \circ p_i^\Sigma = p_\infty^\Sigma \circ \tilde{\rho}_i$ in $B_{s_i}(\mathbf{H}^3/G_i(\Sigma), \tilde{x}_i)$. It follows that for every sufficiently large i, we have $\rho_i \circ p_i^\Sigma|\tilde{\overline{F}}_i = p_\infty^\Sigma \circ \tilde{\rho}_i|\tilde{\overline{F}}_i$ since $\overline{F}_i$ is contained in the K-neighbourhood of x_i. This implies that $\rho_i(\overline{F}_i)$, which we denote by F_i', is lifted homeomorphically to an incompressible surface $\tilde{F}_i'$ in

$\mathbf{H}^3/G'(\Sigma)$, which must be homotopic to $\Psi^{\Sigma}(\tilde{\Sigma})$ as images. (Recall that $\tilde{\Sigma}$ is the homeomorphic lift of Σ to $\mathbf{H}^3/G(\Sigma)$.)

Here the surface F_i' is homotopic to $\Psi(\Sigma)$ as images for sufficiently large i since $\tilde{F}_i'$ is homotopic to $\Psi^{\Sigma}(\tilde{\Sigma})$ since $\mathbf{H}^3/G_i(\Sigma)$ converges to $\mathbf{H}^3/G'(\Sigma)$ by bi-Lipschitz diffeomorphisms as explained before. Furthermore since $\rho_i(\overline{F}_i)$ and $\rho_i(\overline{F}_i^0)$ corresponding to two distinct internal boundary components Σ and Σ^0 are disjoint, so are F_i' and $F_i^{0'}$ corresponding to them. Fix a sufficiently large i, and denoting F_i' by F', we have completed the proof. □

We shall use the symbol $\tilde{F}'$ to denote $\tilde{F}_i' \subset \mathbf{H}^3/G'(\Sigma)$ for the fixed i chosen above. Next we shall consider the case when Σ is an external boundary component

LEMMA 8.15. *Let Σ_j be an external boundary component of N_k. Then there exists an embedded incompressible surface $F'(j)$ in $\mathbf{H}^3/G'$ which is homotopic to $\Psi(\Sigma_j)$ as images. Moreover we can choose such a surface $F'(j)$ so that if $F'(j) \cap F'(j') \neq \emptyset$, then $j = j'$ and $k = k'$, and also $F'(j) \cap F' = \emptyset$ for any F' constructed in Lemma 8.14 from an internal boundary component.*

PROOF. When $\{G_i(\Sigma_j)\}$ converges to a geometrically finite group, i.e, a quasi-Fuchsian group, we can apply the same argument as Lemma 8.14 to get such a surface $F'(j)$.

Suppose that $G_i(\Sigma_j)$ converges to a geometrically infinite group $G'(\Sigma_j)$. Then by Bonahon's theorem in [**5**], $\mathbf{H}^3/G'(\Sigma_j)$ has a simply degenerate end e_j. By Thurston's covering theorem, there exists a neighbourhood $E_j \cong \Sigma_j \times \mathbf{R}$ of e_j such that $p'_{\Sigma_j}|E_j$ is a finite-sheeted covering to its image, where $p'_{\Sigma_j} : \mathbf{H}^3/G'(\Sigma_j) \to \mathbf{H}^3/G'$ is the covering associated with the inclusion. We can prove, as before, that in fact $p'_{\Sigma_j}|E_j$ is an embedding by using the fact that Σ_j is not homotopic to a non-trivial covering of an embedded surface in $\mathbf{H}^3/G$. Then taking a sufficiently large t_j for each j (and k), the surfaces $p'_{\Sigma_j}(\Sigma_j \times \{t_j\})$ satisfy the conclusion of the lemma, where we identify E_j with $\Sigma_j \times \mathbf{R}$. The same argument implies that we can make $F'(j)$ disjoint from the surfaces F' constructed from internal boundary components. □

We define $\tilde{F}'(j)$ to be the homeomorphic lift of $F'(j)$ to $\mathbf{H}^3/G'(\Sigma)$ in this case.

CONTINUATION OF PROOF OF LEMMA 8.8. Let N_0 be a component of $N(G)$ which is not an I-bundle over a closed surface. Let $C(N_0)$ be a compact core of $\mathbf{H}^3/G(N_0)$ which is projected to $N_0 \subset \mathbf{H}^3/G$ homeomorphically by the covering projection. For a boundary component Σ_j of N_0, let $\hat{\Sigma}_j$ be a component of $\partial C(N_0)$ whose image in $\mathbf{H}^3/G$ by the covering projection is Σ_j, and let $\tilde{\Sigma}_j$ be a homeomorphic lift of Σ_j to $\mathbf{H}^3/G(\Sigma_j)$ in accordance with Convention 2. Let $\hat{F}'(j) \subset \mathbf{H}^3/G'(N_0)$ be $p'_{(\Sigma_j,N_0)}(\tilde{F}'(j))$, where $p'_{(\Sigma_j,N_0)} : \mathbf{H}^3/G'(\Sigma_j) \to \mathbf{H}^3/G'(N_0)$ is the converging projection associated with the inclusion, and $\tilde{F}'(j)$ is the surface $\tilde{F}'$ as defined just after the proof of Lemma 8.14 letting $k = 0$ and regarding Σ there as Σ_j or in Lemma 8.15. (Recall that $\tilde{F}'(j) \subset \mathbf{H}^3/G'(\Sigma_j)$ is a homeomorphic lift of $F'(j) \subset \mathbf{H}^3/G'$.) Then $\hat{F}'(j)$ is homotopic to $\Psi^{N_0}(\hat{\Sigma}_j)$ because $\tilde{F}(j)$ is homotopic to $\Psi^{\Sigma_j}(\tilde{\Sigma}_j)$.

Since $\Psi^{N_0}|C(N_0)$ is homotopic to a homeomorphism to a compact core by Lemma 8.4 and $\partial C(N_0) = \sqcup_j \hat{\Sigma}_j$, we can see that $\sqcup_j \hat{F}'(j)$ bounds a compact 3-manifold M_{N_0} which is a compact core of $\mathbf{H}^3/G'(N_0)$.

Let $p'_{N_0} : \mathbf{H}^3/G'(N_0) \to \mathbf{H}^3/G'$ be the covering associated with the inclusion $G'(N_0) \subset G'$. Then $p'_{N_0}|\hat{F}'(j)$ is an embedding whose image is $F'(j)$ given in Lemma 8.14. Since $\sqcup \hat{F}'(j)$ bounds a compact 3-manifold, $\sqcup F'(j)$ is null-homologous after giving an appropriate orientation to each component, hence bounds a compact 3-manifold M' in $\mathbf{H}^3/G'$. As before, let $p_\infty : \mathbf{H}^3/G' \to \mathbf{H}^3/G_\infty$ be the covering associated with the inclusion $G' \subset G_\infty$. Recall that, as was given in the proof of Lemma 8.14 or Lemma 8.15, either $p_\infty(F'(j)) = \rho_i(\overline{F}_i(j))$ for some fixed i and an incompressible surface $\overline{F}_i(j) \subset \mathbf{H}^3/G_i$, or $F'(j)$ is contained in a neighbourhood of a simply-degenerate end. In either case, we can assume that $p_\infty(F'(j))$ and $p_\infty(F'(j'))$ are disjoint unless $j = j'$and $0 = 0'$ by the same argument as in the proof of Lemma 8.15.

By the same reason as before, $\sqcup_j p_\infty(F'(j))$ bounds a compact 3-manifold N_∞ in $\mathbf{H}^3/G_\infty$. Recall that for sufficiently large i', the surface $\rho_{i'}^{-1}(p_\infty(F'(j)))$ is homotopic to $\Psi_{i'}(\Sigma_j)$ since $F'(j)$ is homotopic to $\Psi(\Sigma_j)$. In this situation, since $\rho_{i'}^{-1}(N_\infty)$ is a compact 3-manifold bounded by $\sqcup_j \rho_{i'}^{-1}(p_\infty(F'(j)))$ for a sufficiently large i', it is isotopic to $\Psi_{i'}(N_0)$. Hence N_∞ is homeomorphic to N_0.

Furthermore for any loop γ in N_∞ with the basepoint x_∞ (we can assume that $x_\infty \in N_\infty$ by conjugating the groups G_i if necessary), the loop $\rho_{i'}^{-1}(\gamma)$ can be lifted to a loop in $\mathbf{H}^3/G_{i'}(N_0)$ for sufficiently large i'. Then the lift of $\rho_{i'}^{-1}(\gamma)$ to $\mathbf{H}^3/G_{i'}(N_0)$ converges geometrically to a loop based at the basepoint y_∞ (which is the image of the base point z in the universal cover) in $\mathbf{H}^3/G'(N_0)$ as $i \to \infty$ because $\{G_i(N_0)\}$ converges to $G'(N_0)$ strongly. Hence any loop in N_∞ based at x_∞ is lifted to one in $\mathbf{H}^3/G'(N_0)$ based at y_∞. This implies that N_∞ can be lifted homeomorphically to a compact 3-manifold in $\mathbf{H}^3/G'(N_0)$ homotopic to $\Psi^{N_0}(C(N_0))$, which we denote by $\tilde{N}'_0$. Then $p_\infty \circ p'_{N_0}|\tilde{N}'_0$ is an embedding, which implies that $p'_{N_0}|\tilde{N}'_0$ is also an embedding.

For the case when N_0 is not an I-bundle over a closed surface, it only remains to show that $\tilde{N}'_0$ and $\tilde{N}'_{0'}$ can be chosen so that both they and their projections by p_∞ are mutually disjoint. This can be proved easily by using the facts that $p_\infty(F'^0(j))$ and $p_\infty(F'^{0'}(j'))$ are mutually disjoint unless

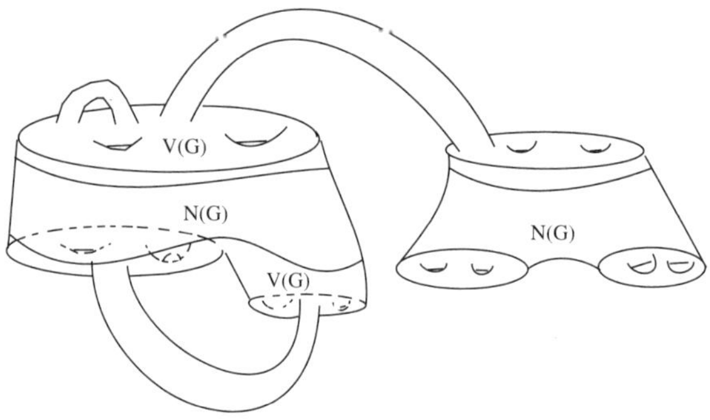

FIGURE 1

$0 = 0'$ and $j = j'$ and that $\rho_{i'}^{-1}(\sqcup_j p_\infty(F'^0(j)))$ and $\rho_{i'}^{-1}(\sqcup_{j'} p_\infty(F'^{0'}(j'))$ bound disjoint compact manifolds in $\mathbf{H}^3/G_{i'}$.

Next suppose that N_0 is a twisted I-bundle over a closed non-orientable surface. Let V_0 be a component of $V(G)$ which contains ∂N_0 as a component of the interior boundary. Recall that we can apply Lemma 8.11 to N_0. Therefore we can see by the same argument as before that there exists an embedded surface $\dot{F}'$ in $\mathbf{H}^3/G'_{V_0}$ which is projected homeomorphically to an embedded surface F' in $\mathbf{H}^3/G'$, which is homotopic to $\Psi(\partial N_0)$. Let $\hat{F}'$ be a homeomorphic lift of F' to $\mathbf{H}^3/G'(N_0)$. Then $\hat{F}'$ bounds a twisted I-bundle $C'(N_0)$ homeomorphic to N_0 in $\mathbf{H}^3/G'(N_0)$. By an argument similar to before, we can choose a $C'(N_0)$ so that it is homotopic to $\Psi^{N_0}(C(N_0))$ and so that $p'_{N_0}|C'(N_0)$ and $p_\infty \circ p'_{N_0}|C'(N_0)$ are embeddings. Let F' be the image $p'_{N_0}(\hat{F}')$, which must be an embedding since its projection in $\mathbf{H}^3/G_\infty$ is an embedding. We can also prove that F' does not intersect the surfaces $F'(j)$ or another F' for a different component of $N(G)$ which is a twisted I-bundle by the same way as before. The same can be proved for the projection $p_\infty(F')$. Hence $C'(N_0)$ and its projection by p_∞ are disjoint from the other components $\tilde{N}'_{0'}$ and its projections in this case. □

8.D. Proof of Lemma 8.2 and Proposition 8.1

PROOF OF LEMMA 8.2. Suppose first that a component N_0 of $N(G)$ is not a product I-bundle over an orientable closed surface. The component N_0 is lifted homeomorphically to a compact core $C(N_0)$ of $\mathbf{H}^3/G(N_0)$. By Lemma 8.8, there exists a compact core $C'(N_0)$ of $\mathbf{H}^3/G'(N_0)$ such that $p'_{N_0}|C'(N_0)$ is an embedding. Further by Lemma 8.4, $\Psi^{N_0}|C(N_0)$ can be homotoped to a homeomorphism to the compact core $C'(N_0)$ without changing Ψ^{N_0} outside an arbitrarily small neighbourhood of $C(N_0)$. Hence $p'_{N_0} \circ \Psi^{N_0}|C(N_0)$ is homotopic to an embedding by a homotopy fixing the complement of an arbitrarily small neighbourhood of $C(N_0)$.

On the other hand, $p'_{N_0} \circ \Psi^{N_0}$ is homotopic to $\Psi \circ p^{N_0}$ where $p^{N_0} : \mathbf{H}^3/G^{N_0} \to \mathbf{H}^3/G$ is the covering associated with the inclusion. Hence Ψ can be homotoped in a small neighbourhood of N_0 so that $\Psi \circ p^{N_0}|C(N_0) = p'_{N_0} \circ \Psi^{N_0}|C(N_0)$ for the Ψ^{N_0} homotoped as above, because the restriction of p^{N_0} to a small neighbourhood of $C(N_0)$ is a homeomorphism to a small neighbourhood of N_0. Thus we have shown that Ψ can be homotoped so that $\Psi|N_0$ is an embedding.

Next suppose that N_0 is a product I-bundle over a closed orientable surface. Let S_0 be an embedded surface in N_0 such that N_0 is regarded as a tubular neighbourhood of S_0. If one of the components of ∂N_0 is contained in $\partial C(G)$, then by the same argument as Lemma 8.15, we can homotope Ψ so that $\Psi|S_0$, hence also $\Psi|N_0$, is an embedding.

When $\partial N_0 \cap \partial C(G) = \emptyset$, there exist two components V_1 an V_2 of $V(G)$, and components S_1 of $\partial_i V_1$ and S_2 of $\partial_i V_2$, such that $S_1 \cup S_2 = \partial N_0$. Then we can apply the same argument for $G(S_0)$ as for $G(\Sigma_j)$ in Lemma 8.14 replacing N_k, which is not an I-bundle over a closed surface, with V_1 to prove that we can homotope Ψ so that $\Psi|S_0$ is an embedding. (See Lemma 8.12.)

By Lemma 8.8, the images of the embeddings $p'_{N_0}|C'(N_0)$ and $p'_{N_{0'}}|C'(N_{0'})$ are disjoint if $0 \neq 0'$ and neither N_0 nor $N_{0'}$ is a product I-bundle over a closed surface. When N_0 is a product I-bundle, the image of the embedding is regarded as a tubular neighbourhood of an embedded surface homotopic to $\Psi|S_0$. As was shown before in Lemmata 8.14, 8.15, and the argument following them in the proof of Lemma 8.8, we can see that such an embedded surface can be taken to be disjoint from the embeddings of the other components of $N(G)$. Hence Ψ can be homotoped so that the image of $\Psi|N_0$ and $\Psi|N_{0'}$ are disjoint if $0 \neq 0'$. This implies that Ψ can be homotoped so that $\Psi|N(G)$ is an embedding.

We can prove the same as above for $p_\infty \circ \Psi|N(G)$ because $p_\infty \circ p'_{N_0}|C'(N_0)$ is an embedding, and $p_\infty \circ p'_{N_0}|C'(N_0)$ and $p_\infty \circ p'_{N_{0'}}|C'(N_{0'})$ are mutually disjoint as was shown in Lemma 8.8 if $0 \neq 0'$ and neither N_0 nor $N_{0'}$ is a product I-bundle over a closed surface. In the case when N_0 is a product I-bundle, which can be regarded as a tubular neighbourhood of an embedded surface S_0, as before, we can use the same argument as the proof of Lemma 8.14 replacing N_k with V_0 or Lemma 8.15, and the argument of the proof of Lemma 8.8 following them to prove that $p_\infty \circ \Psi|S_0$ can be homotoped to an embedding disjoint from the embeddings of the other components of $N(G)$. □

Now we can start the proof of Proposition 8.1.

PROOF OF PROPOSITION 8.1. Recall that the given compact core $C(G)$ of $\mathbf{H}^3/G$ can be constructed from $N(G)$ by adding one-handles. Let $h_1, \ldots, h_s$ be one-handles which are attached to $N(G)$ to form $C(G)$. Let $\alpha_1, \ldots, \alpha_s$ be core arcs of $h_1, \ldots, h_s$ whose endpoints are on $N(G)$. We have already proved that Ψ can be homotoped so that $\Psi|N(G)$ is a homeomorphism to a boundary-irreducible submanifold $N(G')$ of a compact core $C(G')$ in $\mathbf{H}^3/G'$ such that $p_\infty|N(G')$ is an embedding. We shall assume from now on until the end of the proof that $\Psi|N(G)$ is an embedding whose image is equal to $N(G')$.

Consider the arcs $\Psi(\alpha_1), \ldots, \Psi(\alpha_s)$. What we want to do next is to show that we can homotope them fixing the endpoints so that $\Psi|(N(G) \cup (\cup_\sigma \alpha_\sigma))$ is an embedding. For that, it is sufficient to prove for each σ that $\Psi(\alpha_\sigma)$ can be homotoped fixing the endpoints to an arc which does not intersect $N(G')$ at its interior.

Let α_σ^* be the geodesic arc homotopic to $\Psi(\alpha_\sigma)$ relative to the endpoints. Let $\overline{\Psi}_i : \mathbf{H}^3/G \to \mathbf{H}^3/G_i$ be a homeomorphism homotopic to Ψ_i relative to the basepoint, which is guaranteed to exist because (G_i, ψ_i) is a quasi-conformal deformation. We can see that $\rho_i^{-1} \circ p_\infty(N(G'))$, which is homeomorphic to $N(G)$ and we denote by $N(G)_i$, is homotopic to $\overline{\Psi}_i(N(G))$ as images for sufficiently large i by the same argument as before. We can also see that they are also ambient isotopic since they are boundary-irreducible. We isotope $\overline{\Psi}_i$ so that $\overline{\Psi}_i(N(G)) = N(G)_i$. Let α_σ^i be the arc $\rho_i^{-1} \circ p_\infty(\alpha_\sigma^*)$, whose endpoints lie on $N(G)_i$. We can assume that the endpoints of $\overline{\Psi}_i(\alpha_\sigma)$ coincide with those of α_σ^i by isotoping $\overline{\Psi}_i$ preserving the condition that $\overline{\Psi}_i(N(G)) = N(G)_i$.

LEMMA 8.16. *The arc α_σ^i is homotopic to $\overline{\Psi}_i(\alpha_\sigma)$ relative to the endpoints for sufficiently large i.*

PROOF. Let ζ_σ, ξ_σ be the endpoints of α_σ. As $\overline{\Psi}_i(\xi_\sigma)$ stays in a bounded distance from the basepoint x_i as $i \to \infty$ (since $N(G)_i$ converges geometrically to the $p_\infty(N(G'))$ which is compact), we can change the basepoint from x_i to $\xi_\sigma^i = \overline{\Psi}_i(\xi_\sigma)$ by taking conjugates of G_i by bounded elements of $PSL_2\mathbf{C}$ without changing the isometry types of $\mathbf{H}^3/G'$ and $\mathbf{H}^3/G_\infty$. We can regard the conjugation as changing of the basepoints of $\mathbf{H}^3/G'$ and $\mathbf{H}^3/G_\infty$ which keeps the manifolds themselves intact. Then since we defined α_σ^i to be $\rho_i^{-1} \circ p_\infty(\alpha_\sigma^*)$, the points $\{\rho_i(\xi_\sigma^i)\}$ converge to $p_\infty \circ \Psi(\xi_\sigma)$ in $\mathbf{H}^3/G_\infty$, which we shall denote by ξ_σ^∞, and the basepoint of $\mathbf{H}^3/G'$ is equal to $\Psi(\xi_\sigma)$, which we shall denote by ξ'_σ.

Let γ_σ be an essential loop based at ζ_σ. Let γ'_σ be the geodesic loop in $\mathbf{H}^3/G'$ based at ξ'_σ which is homotopic to $\Psi(\gamma_\sigma)$ fixing the basepoint. Then for sufficiently large i, the loop $\rho_i^{-1} \circ p_\infty(\gamma'_\sigma)$ is homotopic to $\overline{\Psi}_i(\gamma_\sigma)$ fixing the basepoint.

Next we change the basepoint of $\mathbf{H}^3/G_i$ again from ξ_σ^i to $\zeta_\sigma^i = \overline{\Psi}_i(\zeta_\sigma)$ by taking a conjugate of G_i corresponding to a move along the arc α_σ^i. Let ζ'_σ and ζ_σ^∞ be the corresponding basepoints of $\mathbf{H}^3/G'$ and $\mathbf{H}^3/G_\infty$ respectively as before. By the way of changing the basepoints, we can see that ζ'_σ coincides with the other endpoint of α_σ^* than ξ'_σ.

Let δ_σ be the loop based at ζ_σ that is equal to $\alpha_\sigma * \gamma_\sigma * \alpha_\sigma^{-1}$, where we regard α_σ as an oriented arc from ζ_σ to ξ_σ. Let δ'_σ be the geodesic loop based at ζ'_σ homotopic to $\Psi(\delta_\sigma)$ fixing the basepoint. Since δ'_σ is homotopic to $\Psi(\alpha_\sigma) * \gamma'_\sigma * \Psi(\alpha_\sigma)^{-1}$, we can see that $\rho_i^{-1} \circ p_\infty(\delta'_\sigma)$ is homotopic to $\alpha^i_\sigma * \gamma^i_\sigma * (\alpha^i_\sigma)^{-1}$ fixing the basepoint for sufficiently large i. On the other hand, similarly to before, we can see that $\rho_i^{-1} \circ p_\infty(\delta'_\sigma)$ is homotopic to $\overline{\Psi}_i(\delta_\sigma) = \overline{\Psi}_i(\alpha_\sigma * \gamma_\sigma * \alpha_\sigma^{-1})$. Thus $\overline{\Psi}_i(\alpha_\sigma * \gamma_\sigma * \alpha_\sigma^{-1})$ is homotopic to $\alpha^i_\sigma * \gamma^i_\sigma * (\alpha^i_\sigma)^{-1}$ fixing the basepoint ζ^i_σ. Since $\overline{\Psi}_i(\gamma_\sigma)$ is homotopic to γ^i_σ fixing the basepoint ξ^i_σ, it follows that the homotopy class of $(\alpha^i_\sigma)^{-1} * \overline{\Psi}_i(\alpha_\sigma)$ commutes with that of γ^i_σ in $\pi_1(\mathbf{H}^3/G_i, \xi^i_\sigma)$.

As the choice of γ_σ was arbitrary, $[(\alpha^i_\sigma)^{-1} * \overline{\Psi}_i(\alpha_\sigma)]$ commutes with every element in the fundamental group $\pi_1(\mathbf{H}^3/G_i, \xi^i_\sigma)$. Thus the only possibility is that $(\alpha^i_\sigma)^{-1} * \overline{\Psi}_i(\alpha_\sigma)$ is null-homotopic fixing the basepoint, which implies that α^i_σ is homotopic to $\overline{\Psi}_i(\alpha_\sigma)$ relative to the endpoints. □

CONTINUATION OF PROOF OF PROPOSITION 8.1. If Ψ can be homotoped so that the arcs $\Psi(\alpha_\sigma)$ are mutually disjoint and their interiors are disjoint from $N(G')$, then $\Psi|C(G)$ is homotopic to an embedding since $C(G)$ is a regular neighbourhood of $N(G) \cup (\cup_\sigma \alpha_\sigma)$. Furthermore the image $\Psi(C(G))$ is then a compact core of $\mathbf{H}^3/G'$ since $C(G)$ is a compact core of $\mathbf{H}^3/G$. In general, we can make the arcs $\Psi(\alpha_\sigma)$ mutually disjoint by perturbing them into general position. Thus what remains to show is that $\Psi(\alpha_\sigma)$ can be homotoped so that its interior is disjoint from $\Psi(N(G)) = N(G')$ fixing the endpoints.

Suppose, seeking a contradiction, that the interior of $\Psi(\alpha_\sigma)$ cannot be homotoped off $N(G')$ fixing the endpoints. We can assume by homotoping Ψ only near α_σ that $\Psi(\alpha_\sigma) \cap \partial N(G')$ is transverse and the number of the components of $\Psi(\alpha_\sigma) \cap N(G')$ is minimum among such arcs homotopic to $\Psi(\alpha_\sigma)$ relative to the endpoints.

The interior of the arc $\Psi(\alpha_\sigma)$ cannot intersect $N(G')$ without intersecting an internal boundary component of $N(G')$, i.e., one which is not homotopic to a boundary component of a compact core, as one can see by the same argument as the proof of Theorem 7.1-(1). (Recall that this is because each boundary component Σ of $N(G)$ homotopic to a boundary component of a compact core, which is incompressible by definition, cuts off a part homeomorphic to $\Sigma \times \mathbf{R}$ from $\mathbf{H}^3/G'$.) Moreover, by the same argument as Lemmata 7.2 and 7.3, one can see that every external boundary component of $N(G)$ is mapped by Ψ to a boundary component of $N(G')$ which is homotopic to a boundary component of $C(G')$. Hence we can assume that the interior of $\Psi(\alpha_\sigma)$ intersects a boundary component of $N(G')$ which is the image by Ψ of an internal boundary component of $N(G)$.

Let $\Sigma' = \Psi(\Sigma)$ be the image of an internal boundary component Σ of $N_0 \sqsubset N(G')$ which $\Psi(\alpha_\sigma)$ intersects. Now by Lemma 8.16, the arc α^i_σ is homotopic to $\overline{\Psi}_i(\alpha_\sigma)$ which is disjoint from $\overline{\Psi}_i(N_0) = \rho_i^{-1} \circ p_\infty(N'_0)$ for sufficiently large i. Since we assumed that $\Psi(\alpha_\sigma)$ intersects Σ', the arc α^i_σ intersects $\overline{\Psi}_i(\Sigma) = \rho_i^{-1} \circ p_\infty(\Sigma')$. Consider a homotopy $H : I \times I \to \mathbf{H}^3/G_i$, which is transverse to $\overline{\Psi}_i(\Sigma)$, such that $H(I \times \{0\}) = \alpha^i_\sigma$ and $H(I \times \{1\}) =$

$\overline{\Psi}_i(\alpha_\sigma)$. Then $H^{-1}(\overline{\Psi}_i(\Sigma))$ must contain a component which is an arc both of whose endpoints lie on $I \times \{0\}$. Hence there must be a subarc β^i_σ of α^i_σ which is homotopic to an arc on $\overline{\Psi}_i(\Sigma)$ fixing the endpoints.

Since $\alpha^i_\sigma = \rho_i^{-1} \circ p_\infty(\alpha^*_\sigma)$, there is an upper bound K independent of i such that length$(\beta^i_\sigma) \leq K$. Let $H' : I \times I \to \mathbf{H}^3/G_i$ be a homotopy between the β^i_σ and the arc on $\overline{\Psi}_i(\Sigma)$. The surface $\overline{\Psi}_i(\Sigma)$ is lifted homeomorphically to a surface $\tilde{\Sigma}_i$ in $\mathbf{H}^3/G_i(\Sigma)$. Since $\{G_i(\Sigma)\}$ converges strongly to a quasi-Fuchsian group by Lemmata 8.6, 8.7, and $\{\overline{\Psi}_i(\Sigma)\}$ converges to $p_\infty \circ \Psi(\Sigma)$ geometrically by our choice of $\overline{\Psi}_i$, there exists a constant K' independent of i such that $\tilde{\Sigma}_i$ is contained in the K'-neighbourhood of the convex core of $\mathbf{H}^3/G_i(\Sigma)$.

The homotopy H' can be lifted to a homotopy between a lift $\tilde{\beta}^i_\sigma$ of β^i_σ and an arc on $\tilde{\Sigma}^j_i$. Since $\tilde{\beta}^i_\sigma$ is contained in the $(K + K')$-neighbourhood of the convex core, a lift of homotopy H' can be chosen to be contained also in the $(K + K')$-neighbourhood of the convex core by constructing homotopy as a union of geodesic arcs. This implies that there exists a constant L independent of i and the homotopy H' can be changed to one whose image is contained in the L-ball centred at the basepoint x_i. Note that such a homotopy can still be lifted to $\mathbf{H}^3/G_i(\Sigma)$. Thus taking sufficiently large i, we can see that the arc $\rho_i(\beta^i_\sigma)$ can be homotoped to an arc on $p_\infty(\Sigma')$ by the homotopy $\rho_i \circ H'$. Moreover, the homotopy can be lifted to one between a subarc of $\Psi(\alpha_\sigma)$, which is a lift of $\rho_i(\beta^i_\sigma)$, and an arc on Σ' fixing the endpoints since H' can be lifted to $\mathbf{H}^3/G_i(\Sigma)$, which converges geometrically to $\mathbf{H}^3/G'(\Sigma)$ covering $\mathbf{H}^3/G'$. By perturbing the part lying on Σ' off $N(G')$ or into $N(G')$, we can reduce the number of the components of $N(G') \cap \Psi(\alpha_\sigma)$. This contradicts our assumption, and the proof of Proposition 8.1 is completed. □

8.E. Concluding the proof of Theorem 2.1

PROOF OF THEOREM 2.1. Finally we can prove our main theorem.

As was remarked before, we can assume that G is not a free group. Let $C(G')$ be a compact core of $\mathbf{H}^3/G'$. Let e be an end of $\mathbf{H}^3/G'$ which is geometrically infinite. There is a component S of $\partial C(G')$ which faces the end e. Let $C(G)$ be a compact core of $\mathbf{H}^3/G$. Then Ψ can be homotoped so that $\Psi|C(G)$ is a homeomorphism onto $C(G')$ by Proposition 8.1. In particular, there is a component S_* of $\partial C(G)$ such that $\Psi|S_*$ is a homeomorphism onto S.

Let Γ be a subgroup of G corresponding to $\iota_\#\pi_1(S_*)$, where ι denotes the inclusion. Then $\Gamma' = \psi(\Gamma)$ corresponds to $\iota_\#\pi_1(S)$. Let V be a component of a characteristic compression body for $C(G)$ containing S_*, and let V' be a component of that for $C(G')$ containing S (if S_* is incompressible, we let V be a collar neighbourhood of S_* in $C(G)$ and V' that of S in $C(G')$.) Then V is lifted homeomorphically to a compact core of $\mathbf{H}^3/\Gamma$ and V' is lifted homeomorphically to a compact core of $\mathbf{H}^3/\Gamma'$.

By the same argument as the special case at the beginning of this section, we can see that either $\Lambda_{\Gamma'} = S^2_\infty$, (which obviously implies that $\Lambda_{G'} = S^2_\infty$,) or there exists a sequence of simple closed curves $\{\tilde{\delta}_j\}$ on the lift $\tilde{S}$ of S whose projective classes converge inside the Masur domain such that the closed geodesic homotopic to $\tilde{\delta}_j$ in $\mathbf{H}^3/\Gamma'$ tends to the end e as $j \to \infty$. We obtain a sequence of simple closed curves $\{\delta_j\}$ on S as in Theorem 4.1 by projecting the sequence $\{\tilde{\delta}_j\}$ because the covering from $\mathbf{H}^3/\Gamma'$ to $\mathbf{H}^3/G'$ associated with the inclusion is proper by the same argument as the proof of Thurston's covering theorem [**45**] (see [**35**]) since we have shown that there is a sequence of pleated surface tending to the end facing the the exterior boundary of the compact core of $\mathbf{H}^3/\Gamma'$ (or we can use the main theorem of Canary [**9**] since we have proved that Γ' is topologically tame.) Because the above holds true for each end e that is geometrically infinite, we have either $\Lambda_{G'} = S^2_\infty$ or the assumption of Theorem 4.1 is satisfied. Thus Theorem 2.1 is reduced to Theorem 4.1. □

Bibliography

[1] L. V. Ahlfors, Finitely generated Kleinian groups, Amer. J. Math. **86**, (1964), 413–423.

[2] L. V. Ahlfors, Fundamental polyhedrons and limit sets of Kleinian groups, Proc. Nat. Acad. Sci. USA **55**, (1966), 251–254.

[3] L. Bers, On boundaries of Teichmüller spaces and on kleinian groups I, Ann. of Math. **91** (1970), 570–600.

[4] F. Bonahon, Cobordism of automorphisms of surfaces, Ann. Sci. Ecole Norm. Sup (4) **16**, (1983), 237–270.

[5] F. Bonahon, Bouts des variétés hyperboliques de dimension 3, Ann. of Math. **124**, (1986), 71–158.

[6] A. Casson and S. Bleiler, Automorphisms of surfaces after Nielsen and Thurston, London Math. Soc. Student Texts **9**, Cambridge Univ. Press, Cambridge, (1988)

[7] R. D. Canary, Ends of hyperbolic 3-manifolds, J. AMS **6**, (1993), 1–35.

[8] R. D. Canary, Hyperbolic structures on 3-manifolds with compressible boundaries, Thesis, Princeton University (1989).

[9] R. D. Canary, A covering theorem for hyperbolic 3-manifold and its applications, Topology **35**, (1996), 751–778.

[10] R. D. Canary, D. B. A. Epstein, P. Green Notes on notes of Thurston, *Analytical and geometric aspects of hyperbolic spaces* London Math. Soc. Lecture Note Ser. **111**, Cambridge Univ. Press (1987), 3–92.

[11] R. D. Canary, Y. Minsky, On limits of tame hyperbolic 3-manifolds, J. Diff. Geom. **43**, (1996), 1–41.

[12] A. Douady, C. J. Earle, Conformally natural extension of homeomorphisms of the circle. Acta Math. **157** (1986), 23–48.

[13] R. Evans, Deformation spaces of hyperbolic 3-manifolds: strong convergence and tameness, PhD Thesis, University of Michigan (2000).

[14] C. D. Feustel, A generalization of Kneser's conjecture, Pacific J. Math. **46**, (1973), 123–130.

[15] D. B. A. Epstein and A. Marden, Convex hulls in hyperbolic space, a theorem of Sullivan, and measured pleated surfaces, *Analytical and geometric aspects of hyperbolic spaces*, London Math. Soc. Lecture Note Ser. **111** Cambridge Univ. Press (1987), 113–253.

[16] A. Fathi, V. Poénaru, et F. Laudenbach, Travaux de Thurston sur les surfaces, Séminaire Orsay, Astérisque **66-67**, (1979)

[17] C. D. Feustel and R. J. Gregorac, On realizing HNN groups in 3-manifolds, Pacific J. Math. **46**, (1973), 381–387.

[18] M. Freedman, J. Hass and P. Scott, Least area incompressible surfaces in 3-manifolds, Invent. Math. **71**, (1983), 609–642.

[19] D. Gabai, Foliations and the topology of 3-manifolds, J. Differential Geom. **18** (1983), 445–503

[20] M. Gromov, Structures métriques pour les variétés riemanniennes, Cedic (1981), Paris

[21] M. Gromov and W. Thurston, Pinching constants for hyperbolic manifolds, Invent. Math. **89**, (1987), 1–12.

[22] J. Hempel, *3-manifolds*, Ann. Math. Studies **86**, Princeton Univ. Press, Princeton New Jersey, (1976).

[23] T. Jørgensen, On discrete groups of Möbius transformations, Amer. J. Math. **98**, (1976), 739–749.

[24] T. Jørgensen and A. Marden, Algebraic and geometric convergence of Kleinian groups, Math. Scand. **66**, (1990), 47–72.

[25] A. Marden, The geometry of finitely generated Kleinian groups, Ann. of Math. **99**, (1974), 465–496

[26] B. Maskit, A characterization of Schottky groups, J. Analyse Math. **19**, (1967), 227–230

[27] B. Maskit, Intersections of component subgroups of Kleinian groups, *Discontinuous groups and Riemann surfaces* Ann Math. Studies **79**, (1974). 349–367

[28] D. McCullough, Compact submanifolds of 3-manifolds with boundary, Quart. J. Math. Oxford **37**, (1986), 299–307

[29] D. McCullough and A. Miller, Homeomorphisms of 3-manifolds with compressible boundary, Memoirs of AMS, **61**-344, (1986)

[30] D. McCullough, A. Miller and G.A. Swarup, Uniqueness of cores of non-compact 3-manifolds, J. London Math. Soc. **32**, (1985), 548–556

[31] W.H. Meeks III, S.-T. Yau, Topology of three dimensional manifolds and the embeddings problems in minimal surface theory, Ann. Math. **112**, (1980), 441–484

[32] J. Milnor, A note on curvature and the fundamental group, J. Diff. Geom. **2**, (1968), 1–7.

[33] J. Morgan, On Thurston's uniformization theorem for three-dimensional manifolds, *The Smith conjecture* Academic Press (1984), 37–125.

[34] K. Ohshika, On limits of quasi-conformal deformations of Kleinian groups Math. Z. **201**, (1989), 167–176.

[35] K. Ohshika, Geometric behaviour of Kleinian groups on boundaries for deformation spaces, Quart. J. Math. Oxford (2), **43**, (1992), 97–111.

[36] K. Ohshika, Strong convergence of Kleinian groups and Carathéodory convergence of domains of discontinuity, Math. Proc. Cambridge Phil. Soc. **112**, (1992), 297–307.

[37] K. Ohshika, Divergent sequences of Kleinian groups, Geometry and Topology Monographs, **1**, The Epstein Birthday Schrift, (1998), 419–450.

[38] J. P. Otal, Courants géodésiques et produits libres, Thèse, Université de Paris-Sud, Orsay.

[39] F. Paulin, Topologie de Gromov équivariante, structures hyperboliques et arbres réels, Invent. Math. **94**, (1988), 353–359.

[40] R. C. Penner with J. L. Harer, Combinatorics of train tracks, Ann. Math. Studies **125**, Princeton Univ. Press, Princeton New Jersey, (1992).

[41] A. Selberg, On discontinuous groups in higher-dimensional symmetric spaces, *Contributions to function theory* (Internat. Colloq. Function Theory), (1960), 147–164 Tata Institute of Fundamental Research, Bombay.

[42] G. P. Scott, Compact submanifolds of 3-manifolds, J. London Math. Soc. **7**, (1973), 246–250

[43] D. Sullivan, On the ergodic theory at infinity of an arbitrary discrete group of hyperbolic motions, *Riemann surfaces and related topics: Proceedings of the 1978 Stony Brook Conference* (State Univ. New York, Stony Brook, N.Y., 1978), Ann. of Math. Stud., **97**, Princeton Univ. Press, Princeton, N.J., (1981), 465–496,

[44] J. Souto, A note on the tameness of hyperbolic 3-manifolds, to appear in Topology

[45] W. Thurston, The geometry and topology of 3-manifolds, lecture notes, Princeton Univ.

[46] W. Thurston, Three-dimensional manifolds, Kleinian groups and hyperbolic geometry, Bull. Amer. Math. Soc. (N.S.) 6 (1982), 357–381.

[47] W. Thurston, On the geometry and dynamics of diffeomorphisms of surfaces. Bull. Amer. Math. Soc. (N.S.) **19** (1988), 417–431.

[48] W. Thurston, Hyperbolic structures on 3-manifolds I : Deformation of acylindrical manifolds, Ann. of Math. **124**, (1986), 203–246.

[49] W. Thurston, Hyperbolic structures on 3-manifolds II : Surface groups and 3-manifolds which fiber over the circle, preprint, (revised version): arXiv: math. GT/9801045 (1998).

[50] W. Thurston, Hyperbolic structures on 3-manifolds III, Deformation of 3-manifolds with incompressible boundary, preprint, (revised version): arXiv: math. GT/9801058 (1998).

[51] W. Thurston, Minimal stretch maps between hyperbolic surfaces, preprint: arXiv:math.GT/980139 (1998).

[52] F. Waldhausen, On irreducible 3-manifolds which are sufficiently large, Ann. of Math. (2) **87** (1968) 56–88.

Index

Editorial Information

To be published in the *Memoirs*, a paper must be correct, new, nontrivial, and significant. Further, it must be well written and of interest to a substantial number of mathematicians. Piecemeal results, such as an inconclusive step toward an unproved major theorem or a minor variation on a known result, are in general not acceptable for publication. Papers appearing in *Memoirs* are generally at least 80 and not more than 200 published pages in length. Papers less than 80 or more than 200 published pages require the approval of the Managing Editor of the Transactions/Memoirs Editorial Board.

As of May 31, 2005, the backlog for this journal was approximately 11 volumes. This estimate is the result of dividing the number of manuscripts for this journal in the Providence office that have not yet gone to the printer on the above date by the average number of monographs per volume over the previous twelve months, reduced by the number of volumes published in four months (the time necessary for preparing a volume for the printer). (There are 6 volumes per year, each containing at least 4 numbers.)

A Consent to Publish and Copyright Agreement is required before a paper will be published in the *Memoirs*. After a paper is accepted for publication, the Providence office will send a Consent to Publish and Copyright Agreement to all authors of the paper. By submitting a paper to the *Memoirs*, authors certify that the results have not been submitted to nor are they under consideration for publication by another journal, conference proceedings, or similar publication.

Information for Authors

Memoirs are printed from camera copy fully prepared by the author. This means that the finished book will look exactly like the copy submitted.

The paper must contain a *descriptive title* and an *abstract* that summarizes the article in language suitable for workers in the general field (algebra, analysis, etc.). The *descriptive title* should be short, but informative; useless or vague phrases such as "some remarks about" or "concerning" should be avoided. The *abstract* should be at least one complete sentence, and at most 300 words. Included with the footnotes to the paper should be the 2000 *Mathematics Subject Classification* representing the primary and secondary subjects of the article. The classifications are accessible from `www.ams.org/msc/`. The list of classifications is also available in print starting with the 1999 annual index of *Mathematical Reviews*. The Mathematics Subject Classification footnote may be followed by a list of *key words and phrases* describing the subject matter of the article and taken from it. Journal abbreviations used in bibliographies are listed in the latest *Mathematical Reviews* annual index. The series abbreviations are also accessible from `www.ams.org/publications/`. To help in preparing and verifying references, the AMS offers MR Lookup, a Reference Tool for Linking, at `www.ams.org/mrlookup/`. When the manuscript is submitted, authors should supply the editor with electronic addresses if available. These will be printed after the postal address at the end of the article.

Electronically prepared manuscripts. The AMS encourages electronically prepared manuscripts, with a strong preference for $\mathcal{A}_{\mathcal{M}}\mathcal{S}$-LaTeX. To this end, the Society has prepared $\mathcal{A}_{\mathcal{M}}\mathcal{S}$-LaTeX author packages for each AMS publication. Author packages include instructions for preparing electronic manuscripts, the *AMS Author Handbook*, samples, and a style file that generates the particular design specifications of that publication series. Though $\mathcal{A}_{\mathcal{M}}\mathcal{S}$-LaTeX is the highly preferred format of TeX, author packages are also available in $\mathcal{A}_{\mathcal{M}}\mathcal{S}$-TeX.

Authors may retrieve an author package from e-MATH starting from `www.ams.org/tex/` or via FTP to `ftp.ams.org` (login as `anonymous`, enter username as password, and type `cd pub/author-info`). The *AMS Author Handbook* and the *Instruction Manual* are available in PDF format following the author packages link from `www.ams.org/tex/`. The author package can be obtained free of charge by sending email to pub@ams.org (Internet) or from the Publication Division, American Mathematical Society, 201 Charles St., Providence, RI 02904, USA. When requesting an author package, please specify $\mathcal{AMS}$-LaTeX or $\mathcal{AMS}$-TeX, Macintosh or IBM (3.5) format, and the publication in which your paper will appear. Please be sure to include your complete mailing address.

Sending electronic files. After acceptance, the source file(s) should be sent to the Providence office (this includes any TeX source file, any graphics files, and the DVI or PostScript file).

Before sending the source file, be sure you have proofread your paper carefully. The files you send must be the EXACT files used to generate the proof copy that was accepted for publication. For all publications, authors are required to send a printed copy of their paper, which exactly matches the copy approved for publication, along with any graphics that will appear in the paper.

TeX files may be submitted by email, FTP, or on diskette. The DVI file(s) and PostScript files should be submitted only by FTP or on diskette unless they are encoded properly to submit through email. (DVI files are binary and PostScript files tend to be very large.)

Electronically prepared manuscripts can be sent via email to pub-submit@ams.org (Internet). The subject line of the message should include the publication code to identify it as a Memoir. TeX source files, DVI files, and PostScript files can be transferred over the Internet by FTP to the Internet node e-math.ams.org (130.44.1.100).

Electronic graphics. Comprehensive instructions on preparing graphics are available at www.ams.org/jourhtml/graphics.html. A few of the major requirements are given here.

Submit files for graphics as EPS (Encapsulated PostScript) files. This includes graphics originated via a graphics application as well as scanned photographs or other computer-generated images. If this is not possible, TIFF files are acceptable as long as they can be opened in Adobe Photoshop or Illustrator. No matter what method was used to produce the graphic, it is necessary to provide a paper copy to the AMS.

Authors using graphics packages for the creation of electronic art should also avoid the use of any lines thinner than 0.5 points in width. Many graphics packages allow the user to specify a "hairline" for a very thin line. Hairlines often look acceptable when proofed on a typical laser printer. However, when produced on a high-resolution laser imagesetter, hairlines become nearly invisible and will be lost entirely in the final printing process.

Screens should be set to values between 15% and 85%. Screens which fall outside of this range are too light or too dark to print correctly. Variations of screens within a graphic should be no less than 10%.

Inquiries. Any inquiries concerning a paper that has been accepted for publication should be sent directly to the Electronic Prepress Department, American Mathematical Society, 201 Charles St., Providence, RI 02904, USA.

Editors

This journal is designed particularly for long research papers, normally at least 80 pages in length, and groups of cognate papers in pure and applied mathematics. Papers intended for publication in the *Memoirs* should be addressed to one of the following editors. In principle the Memoirs welcomes electronic submissions, and some of the editors, those whose names appear below with an asterisk (*), have indicated that they prefer them. However, editors reserve the right to request hard copies after papers have been submitted electronically. Authors are advised to make preliminary email inquiries to editors about whether they are likely to be able to handle submissions in a particular electronic form.

***Algebra** to ALEXANDER KLESHCHEV, Department of Mathematics, University of Oregon, Eugene, OR 97403-1222; email: ams@noether.uoregon.edu

Algebraic geometry to DAN ABRAMOVICH, Department of Mathematics, Brown University, Box 1917, Providence, RI 02912; email: amsedit@math.brown.edu

***Algebraic number theory** to V. KUMAR MURTY, Department of Mathematics, University of Toronto, 100 St. George Street, Toronto, ON M5S 1A1, Canada; email: murty@math.toronto.edu

***Algebraic topology** to ALEJANDRO ADEM, Department of Mathematics, University of British Columbia, Room 121, 1984 Mathematics Road, Vancouver, British Columbia, Canada V6T 1Z2; email: adem@math.ubc.ca

Combinatorics and Lie theory to SERGEY FOMIN, Department of Mathematics, University of Michigan, Ann Arbor, Michigan 48109-1109; email: fomin@umich.edu

Complex analysis and harmonic analysis to ALEXANDER NAGEL, Department of Mathematics, University of Wisconsin, 480 Lincoln Drive, Madison, WI 53706-1313; email: nagel@math.wisc.edu

***Differential geometry and global analysis** to LISA C. JEFFREY, Department of Mathematics, University of Toronto, 100 St. George St., Toronto, ON Canada M5S 3G3; email: jeffrey@math.toronto.edu

Dynamical systems and ergodic theory to ROBERT F. WILLIAMS, Department of Mathematics, University of Texas, Austin, Texas 78712-1082; email: bob@math.utexas.edu

***Functional analysis and operator algebras** to MARIUS DADARLAT, Department of Mathematics, Purdue University, 150 N. University St., West Lafayette, IN 47907-2067; email: mdd@math.purdue.edu

***Geometric analysis** to TOBIAS COLDING, Courant Institute, New York University, 251 Mercer St., New York, NY 10012; email: traneditor@cims.nyu.edu

***Geometric analysis** to MLADEN BESTVINA, Department of Mathematics, University of Utah, 155 South 1400 East, JWB 233, Salt Lake City, Utah 84112-0090; email: bestvina@math.utah.edu

Harmonic analysis, representation theory, and Lie theory to ROBERT J. STANTON, Department of Mathematics, The Ohio State University, 231 West 18th Avenue, Columbus, OH 43210-1174; email: stanton@math.ohio-state.edu

***Logic** to STEFFEN LEMPP, Department of Mathematics, University of Wisconsin, 480 Lincoln Drive, Madison, Wisconsin 53706-1388; email: lempp@math.wisc.edu

Number theory to HAROLD G. DIAMOND, Department of Mathematics, University of Illinois, 1409 W. Green St., Urbana, IL 61801-2917; email: diamond@math.uiuc.edu

***Ordinary differential equations, and applied mathematics** to PETER W. BATES, Department of Mathematics, Michigan State University, East Lansing, MI 48824-1027; email: bates@math.msu.edu

***Partial differential equations** to PATRICIA E. BAUMAN, Department of Mathematics, Purdue University, West Lafayette, IN 47907-1395; email: bauman@math.purdue.edu

***Probability and statistics** to KRZYSZTOF BURDZY, Department of Mathematics, University of Washington, Box 354350, Seattle, Washington 98195-4350; email: burdzy@math.washington.edu

***Real analysis and partial differential equations** to DANIEL TATARU, Department of Mathematics, University of California, Berkeley, Berkeley, CA 94720; email: tataru@math.berkeley.edu

All other communications to the editors should be addressed to the Managing Editor, ROBERT GURALNICK, Department of Mathematics, University of Southern California, Los Angeles, CA 90089-1113; email: guralnic@math.usc.edu.

Titles in This Series

836 **H. G. Dales and A. T.-M. Lau,** The second duals of Beurling algebras, 2005

835 **Kiyoshi Igusa,** Higher complex torsion and the framing principle, 2005

834 **Ken'ichi Ohshika,** Kleinian groups which are limits of geometrically finite groups, 2005

833 **Greg Hjorth and Alexander S. Kechris,** Rigidity theorems for actions of product groups and countable Borel equivalence relations, 2005

832 **Lee Klingler and Lawrence S. Levy,** Representation type of commutative Noetherian rings III: Global wildness and tameness, 2005

831 **K. R. Goodearl and F. Wehrung,** The complete dimension theory of partially ordered systems with equivalence and orthogonality, 2005

830 **Jason Fulman, Peter M. Neumann, and Cheryl E. Praeger,** A generating function approach to the enumeration of matrices in classical groups over finite fields, 2005

829 **S. G. Bobkov and B. Zegarlinski,** Entropy bounds and isoperimetry, 2005

828 **Joel Berman and Paweł M. Idziak,** Generative complexity in algebra, 2005

827 **Trevor A. Welsh,** Fermionic expressions for minimal model Virasoro characters, 2005

826 **Guy Métivier and Kevin Zumbrun,** Large viscous boundary layers for noncharacteristic nonlinear hyperbolic problems, 2005

825 **Yaozhong Hu,** Integral transformations and anticipative calculus for fractional Brownian motions, 2005

824 **Luen-Chau Li and Serge Parmentier,** On dynamical Poisson groupoids I, 2005

823 **Claus Mokler,** An analogue of a reductive algebraic monoid whose unit group is a Kac-Moody group, 2005

822 **Stefano Pigola, Marco Rigoli, and Alberto G. Setti,** Maximum principles on Riemannian manifolds and applications, 2005

821 **Nicole Bopp and Hubert Rubenthaler,** Local zeta functions attached to the minimal spherical series for a class of symmetric spaces, 2005

820 **Vadim A. Kaimanovich and Mikhail Lyubich,** Conformal and harmonic measures on laminations associated with rational maps, 2005

819 **F. Andreatta and E. Z. Goren,** Hilbert modular forms: Mod p and p-adic aspects, 2005

818 **Tom De Medts,** An algebraic structure for Moufang quadrangles, 2005

817 **Javier Fernández de Bobadilla,** Moduli spaces of polynomials in two variables, 2005

816 **Francis Clarke,** Necessary conditions in dynamic optimization, 2005

815 **Martin Bendersky and Donald M. Davis,** V_1-periodic homotopy groups of $SO(n)$, 2004

814 **Johannes Huebschmann,** Kähler spaces, nilpotent orbits, and singular reduction, 2004

813 **Jeff Groah and Blake Temple,** Shock-wave solutions of the Einstein equations with perfect fluid sources: Existence and consistency by a locally inertial Glimm scheme, 2004

812 **Richard D. Canary and Darryl McCullough,** Homotopy equivalences of 3-manifolds and deformation theory of Kleinian groups, 2004

811 **Ottmar Loos and Erhard Neher,** Locally finite root systems, 2004

810 **W. N. Everitt and L. Markus,** Infinite dimensional complex symplectic spaces, 2004

809 **J. T. Cox, D. A. Dawson, and A. Greven,** Mutually catalytic super branching random walks: Large finite systems and renormalization analysis, 2004

808 **Hagen Meltzer,** Exceptional vector bundles, tilting sheaves and tilting complexes for weighted projective lines, 2004

807 **Carlos A. Cabrelli, Christopher Heil, and Ursula M. Molter,** Self-similarity and multiwavelets in higher dimensions, 2004

806 **Spiros A. Argyros and Andreas Tolias,** Methods in the theory of hereditarily indecomposable Banach spaces, 2004